PENGUIN BOOKS

THE BIRD WAY

Jennifer Ackerman has been writing about science and nature for three decades. She is the author of eight books, including the *New York Times* bestseller *The Genius of Birds*, which has been translated into more than twenty-five languages. Her articles and essays have appeared in *Scientific American*, *National Geographic*, *The New York Times*, and many other publications. Ackerman is the recipient of a National Endowment for the Arts Literature Fellowship in Nonfiction, a Bunting Fellowship, and a grant from the Alfred P. Sloan Foundation.

jenniferackermanauthor.com

·   ·   ·

Praise for *The Bird Way*

"[Ackerman's] exhilarating book will leave you as awestruck by the complexities and contradictions of bird life as she is." —*San Francisco Chronicle*

"A revelatory book about the avian world . . . This book is a celebration of the dizzying variety of bird life and behavior, one that will enthrall birders and non-birders alike. . . . It's clear that there's a virtuous circle at work in the scientific study of birds, so that the more we learn about them, the more we recognize the oversimplification and errors of our previous assumptions, the strange and remarkable otherness of life seen through a bird's eye view. *The Bird Way* crystallizes and threads together these revelations into a book full of wonders large and small." —*The Guardian* (London)

"From tales of dazzling plumage to anecdotes about almost unfathomable mimicry, Jennifer Ackerman's *The Bird Way* is a walk through the mysteries, wonders, and peculiarities of the avian world. . . . Ackerman's excitement and love for it are evident in her writing. Her superb storytelling paints a rich picture that engages the reader's imagination, making sometimes-hard-to-grasp research accessible." —*Science*

"In *The Bird Way*, Jennifer Ackerman digs deeper and ranges farther into bird behavior, pulling tasty stories out of rich ground as she hops across the continents. . . . Like a bowerbird, Ms. Ackerman gathers and displays treasures to amaze and delight—then lets the scientists' stories take center stage. . . . Refreshingly, Ackerman spotlights a number of female researchers." —*The Wall Street Journal*

"After reading Ackerman (*The Genius of Birds*), you may listen harder to the various chirps, cheeps, and coos coming from your backyard. Her new book reminds us that we have a lot in common with birds—like us, they are capable of deception and manipulation, not to mention cooperation, culture, and communication."
—*The Washington Post*

"Splendid and spellbinding . . . *The Bird Way* shows us a new way to view birds, yes— but perhaps even better, through their eyes, intellect, and more-than-human senses, it lets birds reveal to us the hidden realities of our shared world."
—Sy Montgomery, *The American Scholar*

"A brilliant synthesis of bird behavior research . . . What makes Ackerman's book a joy to read is not just the stories she tells but her vivid writing style. . . . If there's one thing Ackerman's illuminating book makes clear, it's that there is no *single* way to be a bird. Her opus is a celebration of the sheer diversity of avian behaviors, practices, predilections and the birds she writes about are 'iconoclasts and rule breakers' and remain 'layered in mystery.' It is this decision to focus on birds' idiosyncrasies, to resist generalizations and categorizations, to break down assumptions about bird behavior, and to show how 'individual birds are every bit as distinctive as we humans are' that make this book so remarkable." —*Birding*

"Ackerman brings scientific research alive with personal observations of colorful and fascinating birds, from the kea parrot to the raven to the brush turkey, among others. By showing how each species communicates, plays, parents, works, and thinks, she reminds us that there is no one way to be a bird." —*Science Friday*

"Ackerman packs her book with insightful observations, interesting factoids, and deep dives into new research about birds as varied as seagulls, emus, vultures, and robins."
—*Undark*

"Ackerman's vibrant writing ensures that all things bird are thoroughly compelling and enjoyable." —*Booklist* (starred review)

"Ackerman reminds readers that birds are thinking beings . . . She brings scientific research alive with personal field observations and accounts of her encounters with colorful and fascinating birds. . . . [*The Bird Way*] will engage all readers interested in learning more about birds and natural history." —*Library Journal* (starred review)

"A brightly original book . . . Ackerman is a smooth writer; her presentation of ideas is deft, and her anecdotes are consistently engaging. . . . [She] demonstrates bird science as an evolving discipline that is consistently fascinating, and she offers brilliant discussions of the use of smell, long overlooked but indeed deployed for navigation; courtship signals; predator avoidance, and, not surprisingly, locating food."
—*Kirkus Reviews* (starred review)

The Bird Way

A New Look at How Birds Talk,
Work, Play, Parent, and Think

JENNIFER ACKERMAN

PENGUIN BOOKS

ISBN 9780735223035 (paperback)

THE LIBRARY OF CONGRESS HAS CATALOGED THE
HARDCOVER EDITION AS FOLLOWS:

Names: Ackerman, Jennifer, 1959– author.
Title: The bird way : a new look at how birds talk, work, play, parent,
and think / Jennifer Ackerman.
Description: New York : Penguin Press, 2020. | Includes bibliographical
references and index.
Identifiers: LCCN 2020002317 (print) | LCCN 2020002318 (ebook) |
ISBN 9780735223011 (hardcover) | ISBN 9780735223028 (ebook)
Subjects: LCSH: Birds—Behavior.
Classification: LCC QL698.3 .A284 2020 (print) |
LCC QL698.3 (ebook) | DDC 598.15—dc23
LC record available at https://lccn.loc.gov/2020002317
LC ebook record available at https://lccn.loc.gov/2020002318

Printed in Italy
2nd Printing

Designed by Amanda Dewey

Illustrations by John Burgoyne

The species depicted are as follows, in order of appearance:
white-winged choughs, emperor penguin, canebrake wrens, New Holland
honeyeaters, superb lyrebirds, turkey vultures, black kite, ocellated antbird,
common raven, kea, Arabian babblers, palm cockatoos, spotted bowerbirds,
brush turkey, white-browed scrubwren and fan-tailed cuckoo chick,
greater anis, and common raven.

For Nelle

Contents

LOVE

PARENT

The Bird Way

Introduction

WHEN YOU'VE SEEN
ONE BIRD

There is the mammal way and there is the bird way." This is one scientist's pithy distinction between mammal brains and bird brains: two ways to make a highly intelligent mind.

But the bird way is much more than a unique pattern of brain wiring. It's flight and egg and feathers and song. It's the demure plumage of a mountain thornbill and the extravagant tail feathers of an Indian paradise flycatcher, the solo song of a superb lyrebird and the perfectly timed duets of canebrake wrens, an osprey's hurtling dive toward the sea, and a long-legged heron's still, patient eyeing of the dark water.

There is clearly no single bird way of being but rather a staggering array of species with different looks and lifestyles. In every respect, in plumage, form, song, flight, niche, and behavior, birds vary. It's what we love about them. Diversity fascinates biologists. It fascinates birdwatchers, too, driving us to assemble life lists, to travel to far corners of the globe to visit a rare species or jump in the car to spot a vagrant blown in by a storm, to go "pishing" and whistling into the woods to draw that elusive warbler.

Watch birds for a while, and you see that different species do even the most mundane things in radically different ways. We give a nod to this variety in expressions we use to describe our own extreme behaviors. We are owls or larks, swans or ugly ducklings, hawks or doves, good eggs or bad eggs. We snipe and grouse and cajole, a word that comes from the French root meaning "chatter like a jay." We are dodos or chickens or popinjays or proud as peacocks. We are stool pigeons and sitting ducks. Culture vultures. Vulture capitalists. Lovebirds. An albatross around the neck. Off on a wild goose chase. Cuckoo. We are naked as a jaybird or in full feather. Fully fledged, empty nesters, no spring chicken. We are early birds, jailbirds, rare birds, odd birds.

As biologist E. O. Wilson once said, when you have seen one bird, you have not seen them all.

This is certainly true for behavior. Take white-winged choughs. Australians say it's easy to fall in love with these birds—and it is. They're adorable, charismatic, gregarious, comical: lined up on a narrow tree branch, six or seven red-eyed puffs of black feathers, tenderly preening one another in a pearl-like strand of endearment and affection. Clumsy fliers, they prefer to walk everywhere, swaggering through dry eucalypt woodlands with their heads strutting backward and forward like a chicken's. They pipe and whistle and wag their tails like puppies. They're fond of playing follow-the-leader or keep-away, rolling over one another to win possession of a stick or a slip of bark. About the size of a crow but slimmer—black with elegant white wing patches and an arched bill—they live in stable groups of four to twenty birds and are always, always found in clusters or huddles or lines. Like a tight-knit family, they do everything together, drink, roost, dust bathe, play, run in wide formation like a football team to share a food discovery. Together they build big bizarre nests of mud (or emu or cattle dung if they're in a pinch) set on a horizontal branch, queuing up on the limb, waiting their turn to add their bit of shredded bark, grass, or fur soaked with mud to the rim of the nest. Together they brood, guard, and feed the young. Members of family groups are rarely more than five or ten feet apart. I once saw three fledglings jammed together on the ground like the three wise monkeys, see no evil, hear no evil, speak no evil.

And yet there's a darker side to choughs, especially if the weather turns bad. They squabble and fight, one group pitted against another. Larger groups gang up on smaller groups, flying at them and pecking viciously, dislodging eggs from nests, and nests from trees. They are known to go on violent crime sprees, ruining the nesting efforts of numerous other groups. One bird was observed picking up eggs in its bill one at a time and tossing them to the ground. Perhaps most unsettling, warring choughs do something few animals apart from humans and ants do: They forcibly kidnap and enslave the young from other groups.

This is a book about the range of surprising and sometimes alarming behaviors that birds perform daily, activities that firmly, sometimes gleefully, reverse conventional notions about what is "normal" in birds and what we thought they were capable of.

Lately, scientists have taken a new look at behaviors they have run past for years and dismissed as anomalies or set aside as abiding mysteries. What they have found is upending traditional views of how birds conduct their lives, how they communicate, forage, court, breed, survive. It's also revealing the remarkable strategies and intelligence underlying these activities, abilities we once considered uniquely our own, or at least the sole domain of a few clever mammals—deception, manipulation, cheating, kidnapping, and infanticide, but also ingenious communication between species, cooperation, collaboration, altruism, culture, and play.

Some of these extraordinary behaviors are conundrums that seem to push the edges of, well, birdness: a mother bird that kills her own infant sons, and another that selflessly tends to the young of other birds as if they were her own. Young birds that devote themselves to feeding their siblings, and others so competitive that they'll stab their nest mates to death. Birds that create gorgeous works of art, and birds that wantonly destroy the creations of other birds. Birds like the white-winged chough that contain their own contradictions: one murderous bird that impales its prey on thorns or forked branches but sings so beautifully that composers have devised whole compositions around its songs; another with a reputation for solemnity that is strongly addicted to play; and another that collaborates with one species—humans—but parasitizes another in gruesome fashion. Birds that give gifts and birds that steal, that dance and drum, that paint their creations or paint themselves. Birds that build walls of sound to keep out intruders, and birds that summon playmates with a special call—and may hold the secret to our own penchant for playfulness and the evolution of human laughter.

Earth is home to well over ten thousand different species of birds, many with marvelous, often Seussian, names—the zigzag heron and white-bellied go-away bird, speckled mousebird and naked-faced spiderhunter, the Inaccessible Island rail, pale chanting goshawk, shining sunbeam, military macaw, and wandering tattler, a yellow-legged stanza of elegance I watched probe for crustaceans and worms on the fringes of a tiny island in Alaska's Kachemak Bay. The *wandering* refers to its presence everywhere over vast stretches of sea. *Tattler* refers to the shrill tattling call to alert other birds if an observer approaches too closely. There are whydahs and widowbirds, fantails and fairy-wrens, broadbills and hornbills, and buff-breasted buttonquail (known as BBBQs). Birds live on every continent, in every habitat, even—like the burrowing owl and the Puerto Rican tody—underground. They run to extremes in everything from size and flight style to feather color and physiology. I once saw a biologist weigh a male broad-tailed hummingbird: one-seventh of an ounce. Compare this with the cassowary, a giant weighing one hundred pounds—around twelve thousand times the hummer—that looks as much like a dinosaur as any living bird, can rise up six feet to pluck fruit from limbs, and is capable of killing a man. Or consider the ten-foot wingspan of an Andean condor relative to the five-inch span of a goldcrest.

Some birds are agile fliers, like the northern goshawk, slalom king of the bird world, and swifts and hummingbirds, those avian acrobats. Big, flightless birds, such as the emu and the cassowary, don't take wing at all, although their ancient ancestors did. Likewise, the Galápagos cormorant used to have flight, but lost it over evolutionary time in favor of the grounded life. Seabirds such as the wandering albatross log tens of thousands of miles each year to return to tiny islands in the middle of vast oceans to breed. They may go for years without touching land and, when seas are rough, will sleep on the wing, one eye open to navigate. Bar-tailed godwits migrate from Alaska to New Zealand in a single 7,000-mile flight, traveling day and night for seven to nine days—the longest recorded nonstop migratory flight. In terms of flying distance, the Arctic tern takes all, circling the world in orbit with the seasons. The bird flies from its breeding grounds in Greenland and Iceland to its wintering grounds in Antarctica—a round trip of almost 44,000 miles, the longest migration ever recorded. Over the thirty years of its life, a tern may fly about 1.5 *million* miles, the equivalent of three trips to the moon and back.

As an astronaut who traveled to the International Space Station and made the first all-female space walk in 2019, Jessica Meir knows a thing or two about going to extremes. Meir's goal had always been to walk in space, and on her way to that dream, she explored the lives of two birds capable of truly exceptional physiological feats—one that holds its breath for impossibly long periods of time, the other that flies at breathtaking altitudes.

At Penguin Ranch in Antarctica, Meir investigated emperor penguins, the world's best bird divers. These penguins can dive deeper and longer than any other bird and can tolerate extremely low levels of oxygen in their blood—far below those that would render a human unconscious. Meir observed the birds diving for fish from an underwater viewing chamber. "They look like different animals underwater," she says, "like ballet dancers." The penguins routinely dive for 5 to 12 minutes at a time. One penguin made a 27-minute dive on a single breath. Meir wanted to understand how these animals can stay underwater for so long. "They're air breathers just like we are," she says. "They take a breath before they dive and then use the oxygen in that breath for the entire time they're down there." One of their secrets: They slow their heart rate from 175 beats per minute to around 57 beats per minute, which allows them to slow the use of their oxygen stores.

Later, Meir turned to a bird famous for one of the most extreme migrations on Earth. The bar-headed goose crosses the Himalayas twice a year on its migratory route from sea level in southern Asia up over the enormous mountain range to its summer breeding grounds in the central Asian highlands.

One cold April night in the high Himalayas, naturalist Lawrence Swan stood listening to the silence. From the south a distant sound came, a quiet hum that became a call, the honking of bar-headed geese. Swan followed their movement directly over the summit of Makalu. "At 16,000 feet, where I breathed heavily with every exertion," he writes, "I had witnessed birds flying more than two miles above me, where the oxygen tension is incapable of sustaining human life—and they were calling. It was as if they were ignoring the normal rules of physiology and defying the impossibility of respiration at that height by wasting their breath with honking conversation."

Flapping flight consumes ten to fifteen times more oxygen than resting. Most of these geese reach altitudes of 16,000 to 20,000 feet. One bird was recorded at almost 24,000 feet. At this altitude, oxygen levels are roughly a half to a third what they are at sea level. Bar-headed geese sustain the high oxygen demands of flight in air that is so thin that even the most elite human athletes can barely walk in it.

Meir wondered, do the birds use thermals, those upward currents of warm air, to save energy? "No, they actually fly at night and early in the morning, when there's a strong headwind and the temperature is lower," she says. Moreover, they're flapping fliers and almost never glide or soar. So how do they do it?

To find out, Meir decided to train them to fly in a wind tunnel. And to do so, she became Mother Goose, raising a gaggle of twelve goslings from birth so that they would imprint on her. "We would go for walks together, take naps together," she says. "It's true what they say about children, they grow up fast." She started the geese flying by riding a bike so that they would fly right next to her, almost beak to cheek. That worked for a day, but they were too fast, so she took to riding a motorcycle up and down small roads, the birds by her side, wingtips brushing her shoulders. "Looking into the eye of a bird like that is really special," she says. Eventually, Meir and her colleague Julia York of the University of Texas readied the geese for flight in the wind tunnel, fitting them with tiny backpacks that recorded their vital signs and with special custommade masks that changed the oxygen content of the air they breathed to mimic those they might experience through the passes of the Himalayas and at the summit of Mount Everest. Then they set the birds flying in the tunnel to measure their heart rate, metabolic rate, blood oxygen levels, and temperature under different conditions.

Scientists knew that these geese have several adaptations that help them at high altitudes: larger lungs than other birds, more efficient breathing (deeper and less frequent), a kind of hemoglobin that grabs oxygen more effectively (allowing them to extract more of the gas from each breath of air than can other birds), and blood capillaries that are especially densely distributed throughout their muscles to deliver the oxygen. What Meir and York learned through their experiments was that the geese have yet another superbird mechanism: a unique response to temperature. In their bodies, the temperature difference between their cold lungs and their warm muscles can increase the delivery of oxygen by twofold during sustained flapping flight at high altitudes. The geese also minimize their metabolic rate, reducing the amount of oxygen they need to fly.

"But this isn't the whole story," says Meir. "We still don't know how these birds cope with the low barometric pressure at extremely high altitudes, which would do in other species."

It's what I love about so many aspects of bird biology and behavior. They're still layered with mystery.

Then there's the vast spectrum of plumage in the bird world, a riot of brilliantly hued buntings and carnival-colored parrots; the vibrant Palawan peacock-pheasant, its glossy blue-black feathers lustered with a dazzling metallic green; the red bird-of-paradise, with its filmy plumes and long plastic-like feather wires projecting from its tail, and its cousin, the paradise riflebird, with its outlandish super-black feathers created by unusual bristling microstructures that trap nearly all light; as well as the whiskered auklet of the Aleutian Islands, which sprouts acutely sensitive plumes from its head that guide it through its dark nest cavities in the nesting season.

James Dale studies color in birds and how they use it. "Birds can't use their color as a weapon, but they can use it to avoid conflict," he says. An ornithologist from New Zealand (land of the bright purple pukeko), Dale has devoted his career to making sense of the fantastic variety. There are some rules, he told me. Three in particular: Males are flashier than females, which are often a dull color so that they blend in with their surroundings while they're incubating eggs. Adults are more colorful than youngsters. Birds are brighter in the breeding season.

"But birds are rule breakers," he says. To name a few renegades: Female red phalaropes and painted snipes are more colorful by far than the muted males of the species. American coot chicks, with their bright red beaks and caps, outshine their dull parents—and for very good reason. Parent coots preferentially feed ornamented chicks over non-ornamented chicks. In male red-backed fairy-wrens, it's the social environment that determines whether young males molt into their flashy red-black breeding plumage—specifically, whether there are any old males around to harass the young birds and drive them away.

Perhaps chief among the color rebels is a parrot that lives in remote areas of northern Australia and New Guinea, *Eclectus roratus* (from the same Greek root as *eclectic*, and *roratus* for the sheen on the bird's plumage).

"Few birds have puzzled scientists more than this parrot," says Robert Heinsohn, a professor of evolutionary and conservation biology at Australian National University who studied the bird for nearly a decade. Heinsohn reports that when the great evolutionary biologist William Hamilton gave lectures, he would show an image of a male and female eclectus sitting together. The male was a bright grassy green and the female a resplendent crimson, her belly "bedewed with a blue haze," as the bird's European discoverer described it—in stark contrast to the normal pattern in dimorphic birds, where females are drab and males are brightly colored. "No other bird has both sexes so 'beautified' in different ways," says Heinsohn. In fact, so flamboyant is the female's plumage and so distinct from the male's that for the first hundred years after the parrots were discovered, people thought they were separate species. "Then one day," says Heinsohn "some naturalist saw a green one on top of a red one."

In a handful of other species, females sport brighter, fancier plumage than males. These include phalaropes, spotted sandpipers, painted snipes, wattled jacanas, and button quail. But in each of these cases, there is a reversal of usual sex roles, with males incubating eggs and females defending territories and fighting among themselves for access to males. "So these species are really the exceptions that prove the rule because they demonstrate that the competitive sex is the one most likely to have bright colors," says Heinsohn.

Not so among the radical eclectus parrots. There's no role reversal here: Females incubate eggs and raise their young. Moreover, even the chicks are rule busters. Unlike the chicks of most birds, which hang on to their drab unisexual juvenile plumage for at least the first year of their life, eclectus chicks hatch with sex-specific down colors and then molt straight into the dramatic full-color adult plumage.

According to Heinsohn, William Hamilton ended his lecture featuring the parrots with the line "When I understand why one sex is red and the other green, I will be ready to die." Sadly, Hamilton died of malaria contracted on an expedition to the Congo before Heinsohn unraveled this mystery—and another, perhaps even stranger, puzzle that's closely knit to it.

If the feathers of eclectus parrots are odd, their breeding conduct is even stranger. Female eclectus parrots are known to kill their own sons as soon as they hatch. This is one of those behavioral riddles so counterintuitive it boggles the mind.

From a biological point of view, it's easier to understand infanticide when it involves killing the young of others for food or other competitive reasons. But killing your own offspring? Producing young is so energetically expensive that producing them and then promptly doing away with them makes little biological sense.

It's even harder to grasp why a parent would systematically kill off just one sex. This sort of sex-specific infanticide is extremely rare in the animal world. Aside from the wasted effort, it leads to inequities in the sex ratio of a population: too many females competing for too few male mates or the other way around. But as Heinsohn discovered over ten years of research in remote northern Australia, mother eclectus birds sometimes target the removal of their male nestlings within three days of hatching. Heinsohn often found the chicks at the base of a nest tree, pecked to death.

Why would a mother kill her sons? What drives a bird to such extreme behavior? And what possible value could it have to her reproductive survival?

Birds display plenty of behaviors at the more altruistic end of the spectrum—helping, cooperating, collaborating, acting selflessly. One example: the lance-tailed manakin's tightly choreographed cooperative display of two males performing fluttering, twitching somersaults to woo females. Only one of the males, the alpha, gets to mate; the beta male is always relegated to the role of wingman, and yet time after time, he pours his heart into his best possible performance. Some birds raise young that are not their own, bestowing on them the same devoted parental attention and nurturing they would give to their own offspring. Bald ibises work cooperatively when they migrate, gamely taking turns leading and following the V formation, precisely matching the amount of time in lead and trailing positions. Kea, those brainy, playful parrots from New Zealand, collaborate on tasks in a way we previously thought was possible only in humans.

Even individual birds within a species have their own idiosyncrasies. Watch a murmuration of starlings or thousands of birds in a seabird colony—like the nesting kittiwakes I saw on Gull Island in Kachemak Bay one May day, all keening, swirling, and so apparently of one mind they seemed a single organism rather than fourteen thousand—and it's easy to assume that all members of a single species act alike. Indeed, it was thought for years that birds of a kind responded in the same way to a given situation with a sort of stereotyped behavior or fixed-action pattern. But naturalists and scientists who spend long hours attentively observing birds and live intimately with them often learn to recognize individuals by their unique personalities, characteristic mannerisms, telltale behaviors, even their distinguishing faces.

Birds certainly recognize one another as individuals. Precocial chicks like goslings and ducklings that follow their parents just hours after hatching learn to recognize particular adults at a surprisingly tender age, by look, voice, personality. Seabirds can often recognize their flying mates from a distance. Many birds can spot their neighbors as individuals and may be sociable toward some and antagonistic toward others.

While there are trademark behaviors that one can use to identify species—the teeter-bob of a spotted sandpiper, for instance—individual birds are every bit as distinctive as we humans are. Members of a species may share basic dance steps, but each bird is a ballerina with her own unique style of moving, foraging, talking, courting, mating. "If one wishes to understand the behavior of animals," writes zoologist Donald Griffin "one must take account of their individuality, annoying as this may be to those who prefer the tidiness of physics, chemistry, and mathematical formulations."

This book explores five arenas of daily activity for birds—talk, work, play, love, and parenting—and tells the stories of extreme examples. For instance, the elaborate "talk" of two different birds, one that packs its sentences with far more meaning than we ever thought possible, all for the general good, and one that speaks fluently in foreign tongues to manipulate and deceive others for selfish purposes. Both stories illuminate deep mysteries in bird communication and reveal its subtle, language-like qualities. The book also looks at the surprising range of ways birds raise their young, from the zero-effort parenting of brood parasites that slip their eggs into the nests of other bird species and leave all nurturing of their young to these stranger-hosts—a subversive act that turns out to require some highly sophisticated intelligence—to the other extreme, the communal parenting of greater anis of Panama, which coordinate efforts to collectively rear chicks in egalitarian helper groups of up to a dozen birds.

Why focus on extreme behaviors? "Odd examples of behavior are always revealing," says Robert Heinsohn. "Sometimes they provide a strong contrast to what usually happens, the exceptions that prove the rule, offering insights and perspectives on what's typical in the bird world." Other times the behaviors are just out there teaching us to think about birds in a new way. "It's like taking everything in the room and turning it ninety degrees," says Heinsohn. "Suddenly you see a new picture." We've learned not to ignore the outliers. They often have something important to tell us about what it takes to succeed as a bird, especially under difficult circumstances. Unusual behavior in birds is often an indicator of ingenious adaptation to difficult problems or dire environmental conditions.

A menagerie of species makes a showing here, from vultures to veeries, cranes to canebrake wrens. Some birds appear and reappear. Hummingbirds, for instance. Anyone who has met these little birds knows they're extreme, a ton of truculence packed into a feathered fraction of an ounce. Fiercely territorial, they behave like Chihuahuas that think they're mastiffs. There's good evidence that, in some settings at least, they act like sociopaths.

Australian species crop up throughout the book. There's a reason for that. As biologist Tim Low writes in his brilliant book *Where Song Began*, "Extreme behavior in birds is more likely in Australia than anywhere else." Australian birds occupy more ecological niches than birds anywhere on earth. They tend to be longer lived and more intelligent than birds on other continents. Also, Australia is where some fundamental aspects of bird being were born. Like song.

I spent six weeks on the southern continent, tagging along with Low and other Australian naturalists and scientists studying odd bird behaviors. Australia is a continent populated by outlandish creatures that almost defy description—kangaroos, platypuses, wombats, swamp wallabies, water dragons—and landscapes straight out of Arcadia, with spiked palms, bottlebrush blossoms, yellow wattles and blue gums, and flame trees bursting with blooms almost impossibly red. But I was most besotted with the birds.

When the English ornithologist John Gould visited Australia in the mid-nineteenth century, he observed that the great southern land contained bird "peculiarities unexampled in any other portion of the globe." The birds there were *striking, remarkable, extraordinary*, and *unrivalled*. Well, most especially one group of birds that seemed to defy all traditional assumptions about avian behavior. Here were creatures who built bowers, what he called "playing-places," which they spent hours meticulously decorating with profuse treasures, arranged by color and likeness, each according to the fancy of its species. (The wondrousness of these birds did not stop Gould from shooting, skinning, and eating them.) But any number of Australian birds could earn his superlatives. The palm cockatoo, for instance, a bird with a huge hooked bill and spray of dark crest feathers that literally crafts its own musical instruments. Or the megapodes that build giant mounds up to fifteen feet high and bury their eggs there so their young must kick their way upward from beneath tons of debris. Or the lyrebirds, finest vocalists of the bird world, that sing their tails off in winter. There are currawongs, butcherbirds, lorikeets, birds-of-paradise, and everywhere, everywhere, Australian magpies, loud, intelligent, often combative birds, known to launch vicious attacks on other species including humans when provoked. During the nesting season, you see bikers riding around wearing helmets festooned with elaborate forests of pipe cleaners or party poppers to deter the swooping birds. Australians often seem oblivious of this gorgeous parade of big, bold, *bizarre* birds around them—the galahs, common as starlings but painted delicate pink, and the raucous sulphur-crested cockatoos, with their shattering screeches and fancy upswept yellow crests. It was a cockatoo named Snowball that recently gained fame for his ability to choreograph his own dance steps in time to Queen and Cyndi Lauper—fourteen distinct moves, from head bobs and foot lifts, to body rolls and the Madonna vogue—suggesting, as the researchers said, that "spontaneity and diversity of movement to music are not uniquely human."

But extreme bird behavior is not limited to the great southern continent. Numerically, Central and South America have by far the greatest diversity of bird species, and plenty of them display roguish behavior that gives Australian renegades a run for their money. For instance, the long-tailed hermit hummingbird of Venezuela and the Guianas, a bird that impersonates other, competing males and then murders them to take their place in a mating arena. Or Brazil's white bellbird, the world's loudest bird, with its piercing, musical two-tone gonging—louder than the bellow of a bison or the howling of a howler monkey—which it uses to woo a mate. Or the ocellated antbirds found in Central America and northern Ecuador that have thoroughly learned the ways of another class of animals—ants—mastering their habits through methods of learning, memory, and information sharing we thought possible in only a handful of species, including ours.

The idea for this book was seeded in conversations about novel bird behaviors with Louis Lefebvre of McGill University during research for my last book, *The Genius of Birds*. More than two decades ago, Lefebvre invented the first scale of intelligence for birds, based on a bird's behavior in the wild. How inventive is the species in its natural environment? Does it make use of new things and find creative solutions to the problems it faces? Does it try new foods? These activities are indicators of what's called behavioral flexibility, which is one fairly reliable measure of intelligence. It's the ability to do something new—to change your behavior to address new circumstances and new challenges. Ornithological journals are full of short reports of these kinds of odd and interesting doings. Lefebvre had combed through journals of the past seventy-five years and found more than two thousand reports of these sorts of innovative behaviors in birds of different species. A prime example was the hooded crows that stole fish from ice fishermen by tugging on their lines with their beaks and walking across the ice as far as they could go, then returning for another stretch of line, stepping on it each time to make sure it didn't slip back.

A recent, more high-tech instance of bird ingenuity popped up in 2018 when a scientist tracking western gulls with geolocators to see where they fed was puzzled to see a gull traveling at sixty miles per hour for a distance of seventy-five miles, crossing the Bay Bridge from San Francisco to Oakland and traveling along the interstates before returning by the same route to her nest. It turned out that the gull, a female breeding on the Farallon Islands west of San Francisco Bay, had hitched a ride on a garbage truck bound for an organic composting facility in the Central Valley near Modesto. At first the researcher thought the bird might have gotten trapped in the truck. But then, two days later, the same thing happened. Clearly, this gull was using its head (if not its palate—as one Bay Area news reporter quipped, "It might be the only time a San Francisco resident ever drove to Modesto for dinner").

Scientists traditionally have little use for anecdotal evidence, demanding data that can be replicated or manipulated statistically. But a single observation by a competent and honest observer of a bird doing something exceptional can offer a rare window into a bird's flexibility of mind. The reports are anecdotal, to be sure, but together they produce plentiful evidence of the ability of birds to solve problems or discover new and better ways to accomplish daily tasks.

The point is this: Novel or unusual behavior is often intelligent behavior.

When I asked scientists from all over the world for examples of striking bird behaviors in the wild, again and again they led me to stories of ingenuity and cleverness—smart strategies, sometimes rooted in evolutionary wisdom, but more often based in a bird's capacity for complex cognition. That's broadly defined as the ability to acquire, process, store, and use information in different contexts. In the past decade or so, birds have revealed their ability to solve problems using advanced cognitive skills rather than simple instinct or conditioning, learning by association. These sophisticated mental skills—such as decision-making, finding patterns, and planning for the future—are what allow birds to flexibly fine-tune their behavior in response to challenges of all kinds over their lifetimes.

Only lately has science illuminated how birds can be smart with a brain at best the size of a walnut. In 2016, a team of international scientists reported their discovery of one secret: birds pack more brain cells into a smaller space. When the team counted the number of neurons in the brains of twenty-eight different bird species ranging in size from the pint-size zebra finch to the six-foot-tall emu, they found that birds have higher neuron counts in their small brains than do mammals or even primates of similar brain size. Neurons in bird brains are much smaller, more numerous, and more densely packed than those in mammalian and primate brains. This tight arrangement of neurons makes for efficient high-speed sensory and nervous systems. In other words, say the researchers, bird brains have the potential to provide much higher cognitive clout per pound than do mammalian brains.

Moreover, says neuroscientist Suzana Herculano-Houzel, who led the research, in the brains of parrots and songbirds, most of the "extra" neurons occur in the pallium region of the forebrain, the part of the bird brain that corresponds to our cerebral cortex and is typically associated with intelligent behavior. In fact, big parrots like macaws and cockatoos, as well as corvids such as ravens and crows, have higher neuron counts in the forebrain than do monkeys with much larger brains—in some cases, twice as many neurons, with more connections between them—which explains why these birds are capable of cognitive feats comparable to those of great apes.

Birds have shown us a different way to shape an intelligent brain. Mammals use larger neurons to connect distant brain regions; birds keep their neurons small, close together, and locally connected and grow only a limited number of larger neurons to handle long-distance communication. In building powerful brains, says Herculano-Houzel, nature has two strategies: It can tinker with the number of neurons and their size, and also, it can change their distribution in different parts of the brain. In birds, nature uses both tactics—to brilliant effect.

Exploration of curious behaviors in birds is overturning some fundamental beliefs about them. Take song. Ornithologists in the Northern Hemisphere have traditionally considered complex birdsong to be an almost exclusively male trait and have tended to dismiss instances of female song as rare or atypical. In the past few years, a closer look has toppled this view. Female song is no anomaly or aberration but widespread in songbirds, especially in species that live in tropical and subtropical regions, but also in temperate regions.

Many behaviors once thought simple and straightforward are, on second glance, turning out to be far more nuanced and complex than once imagined. Mating systems, for instance. Some breeding arrangements in birds once believed to be a simple matter of monogamous pairings are in fact exceeded in complexity only by those of humans. The way some birds forage has less to do with the keen eyesight we thought was critical for food finding and more with a sophisticated hound-like nose. The apparently uncomplicated alarm calls birds make in response to threats are packed with more meaning than we ever thought possible—and fully grasped by many different species, not just the tribe of the alerting bird. Some birds, it seems, have developed a kind of universal language.

W hy are these surprising insights turning up now?

For one thing, scientists are shedding biases that have blinkered research for generations. Sensory prejudices, for instance—the notion that the world we humans see, hear, and smell is the world experienced by other creatures. In fact, it's strictly our reality, constrained by our cognitive, biological, even cultural limitations. Other animals experience other realities. The human sensory bias has sometimes blinded us to the differences in bird sensory capacities—and their diversity. But new ways of studying bird perception has helped us take a bird's-eye view of the world, changing the way we see what they see and revealing the hidden layers of their reality—how they see unimaginable explosions of color and pattern, hear sounds inaudible to our ears, smell the shape of whole landscapes.

Then there's geographic bias. We thought we had a handle on the bird way of doing things based on bird behavior in the Northern Hemisphere, primarily northern North America and Europe. That's where most ornithologists were working until quite recently. A few species of ducks harvested by hunters in the north were far better studied than the myriad small-winged natives of the neotropical rainforest canopy. For decades, temperate birds set the norm: Group breeding is exceedingly rare. Migration is typical. Only males sing complex songs and mostly in the breeding season. Only songbirds can see ultraviolet light. The relationship between brood parasites and the birds they target is a neat and tidy evolutionary arms race between a single parasite and a single host.

None of this is so. Birds in temperate regions, it turns out, are often the exceptions rather than the rule. Many of their habits and behaviors are typical primarily of birds with a short breeding season and birds that migrate—a relatively new development from an evolutionary standpoint. And their way of singing—males advertising their territory for a brief period of breeding—is specialized and atypical of the bird world at large. Now that scientists are focusing more on tropical species, the northern blinders are lifting off, and a new vision of what's usual and unusual in the bird world is beginning to emerge.

Ornithological vision has been skewed not just by the hemispheric bias of researchers but also by their gender and gender bias. Until quite recently, most ornithologists were men, and research tended to focus on what male birds were up to; the part played by female birds in the life histories of their species, from female ornamental traits to breeding systems, was often downplayed or ignored.

In 2016, at the biggest ornithology meeting ever convened in Washington, DC, a group of researchers met for a roundtable discussion of birdsong. It was led by Karan Odom and Lauryn Benedict, who with a team of international scientists had made recent discoveries that were overturning the long-held theory that complex birdsong was almost exclusively a male trait. Benedict told the story of when she was doing fieldwork as a graduate student, and she and her fellow researchers heard female birds "making strange noises, singing, and doing other cool vocal stuff we didn't understand." But they weren't publishing their findings, because they thought this was just aberrant behavior in species that had already been thoroughly studied by male ornithologists.

Thanks to Odom and Benedict and women scientists like them, the world of ornithology is changing. When the pair asked the researchers assembled at the table for their observations, one example after another of female song popped up around the circle: female prothonotary warblers with unique songs they use to win mates in the early stages of courtship; female Florida scrub jays singing warble songs. Dustin Reichard, a North American researcher who called himself "a recovering female song denier," had noticed singing in female dark-eyed juncos in his own study population.

New tools are also changing the game, among them, new technologies for observing birds in the wild, tracking their movements over short and long distances, and monitoring their behavior. Tiny backpacks loaded with special devices attached to the heads of frigate birds, for instance, revealed some surprising sleep patterns. The birds doze while in flight, usually one brain hemisphere at a time, but they also fall into whole-brain sleep—just for a few seconds at a time—a quick in-flight power nap.

Webcams and miniature video cameras are offering a close look at behaviors normally hidden or occurring at such speed they're too quick for our eyes. The world of birds moves about ten times faster than ours, and only with high-speed video can we see some of their amazing feats: tap dancing to a beat, turning somersaults in the air, executing display moves as complex, coordinated, and beautiful as those of any gymnast.

Molecular tools, too, have sharpened our vision. DNA analysis revolutionized our understanding of bird origins and evolution—showing, for instance, that all our beloved songbirds of the Northern Hemisphere trace their origins to ancestors living in Australasia and New Guinea forty-five million to sixty-five million years ago. Molecular fingerprinting has transformed our view of bird relationships, putting to rest the myth of birds as champions of monogamy and revealing surprising alliances between birds that are not kin, just good collaborators.

Breakthroughs have also come from the study of wild cognition, the sophisticated ways birds learn and solve problems in their natural settings. Not long ago, scientists largely limited their study of bird cognition to the laboratory, where they had strict control over any testing conditions that might affect a bird's performance—sights, sounds, smells, lighting, temperature, the presence of other birds, as well as the bird's internal state, its hunger, and its prior experience. "In those early days, avian cognition used to be pretty much about doing something with a pigeon in a box," says Sue Healy of the University of St. Andrews. This was—and is—a useful way of investigating what birds can learn and remember. It taught us about the impressive visual and memory capacities of pigeons, for instance: In a laboratory setting, pigeons can remember hundreds of images for longer than a year. And because of their ability to make subtle visual distinctions, they have even been trained to detect the difference between normal and cancerous tissue in mammograms—with more accuracy than a trained technician. But how do they use this ability in their day-to-day life?

While some birds like pigeons and zebra finches are naturals in the lab, unfazed by human-made environments and devices, others don't take well to an artificial setting and don't reveal their true capacities in an experimental setup. Test the memory of a coal or marsh tit with a touchscreen computer in a lab, and it performs miserably, holding in mind an image for at most a few minutes—whereas in the field, it can remember the locations of individual food caches for months.

In field studies on the cognition involved in nest building in birds in the wild, Healy and her colleagues have gleaned insights into the complexity of what was once considered a simple, hardwired behavior. They have discovered that blue tits know enough about the weather and its effect on their young to build different nests at different temperatures, and that the diverse kinds of nesting structures built in different white-browed sparrow-weaver colonies are the upshot of social learning—one bird observing and learning from another.

Healy's research on the cognitive abilities of foraging rufous hummingbirds in the wild has revealed the astonishing memories of these tiny birds. With a brain the size of a grain of rice, they can keep a running tab of multiple aspects of their visits to flowers—which ones offer the best-quality food, how quickly they'll refill with nectar, and when they are worth revisiting—displaying a type of memory once attributed solely to humans.

"Testing cognition in the wild is difficult," says Healy. "There's a reason why nearly everything we know comes from pigeons. But working with these smart species is really satisfying. It can take two years to train a pigeon; a hummingbird we can train in a day." It's not that field studies are better than lab studies, Healy says; they're just different. "We can't do in the field all of the beautiful manipulations we can do in the lab, but we can see what birds do in the open environment."

A few kinds of birds can make and use their own tools in the lab, including vasa parrots and Goffin's cockatoos. But do they actually do so in the wild? "A big advantage to testing birds in the field is the possibility of seeing not just what they *can* do," says Healy, "but what they actually *do* do when faced with social and ecological challenges."

"It's an exciting time to be studying bird behavior," Healy says. In one arena after another, birds are revealing the secret, sophisticated intelligence underlying their natural—and sometimes seemingly *unnatural*—behaviors and showing us how consistently we have underestimated what's going on in their minds. It's clear that birds are thinking beings, even if they're thinking about different things, in different ways, than we humans do.

Birds are iconoclasts and rule breakers. They destroy our assumptions. They defy our neat categories and tidy unifying theories that try to explain all the mystifying variety under one big umbrella. They blow apart our beliefs about the uniqueness of our own species. Time and again we humans have claimed that we're the only species with a particular capacity—toolmaking, reasoning, language-like communication—only to discover that birds share similar abilities. The more we learn about the range of their extraordinary behaviors, the more birds defeat our efforts to pigeonhole them, if you'll excuse the expression.

Talk

Chapter One

DAWN CHORUS

I once stood on the edge of a salt marsh near Kachemak Bay, watching sandhill cranes dip and bow, ruffle and strut. I had no idea what I was looking at until I consulted a handy little dictionary of crane displays and postures created by George Happ and Christy Yuncker after years of field research. The ruffle of feather tuft, the expansion of a red skin patch, the raising or lowering of the head. These are *embodied* words, subtle or confusing to our eye, but to the birds are clear as day, expressing emotion, conveying intent, signaling social purpose. Head and neck "craning" forward means "ready to fly," a signal to the family. Head lifted high at right angles to erect neck, bare red skin expanded, means "on alert," watching for potential threat. Feather tuft on top of head sleeked down: mild arousal. Crouching, spreading wings and drooping them to the ground: a rare high-intensity show of aggression, usually by a female.

Sandhill cranes intentionally paint their own feathers, using tufts of muddy grass to daub them with iron-rich red mud, possibly for camouflage or to repel insects. Other birds, such as herons, pelicans, and ibis, use so-called cosmetic coloration for sexual signaling. Perhaps the most spectacular example occurs in the endangered toki, or Japanese crested ibis, which applies to its white feathers a black, greasy secretion that oozes from its head and neck as nuptial coloration during the breeding season.

Birds are the great communicators of the animal world. They talk while they court and while they fight, while they forage and while they travel, while they stave off predators and while they raise their young. They speak with their voices, their bodies, and their feathers. They may not have the facial musculature we primates use to express ourselves, but they can powerfully communicate their inner states with head and body, with facial feathers, crests, gestures, displays of wings and tail, as sandhill cranes do.

Take a gander at the graphic chatter of the red-billed quelea of sub-Saharan Africa, a bird that nearly bowled over the researcher who discovered its unusual form of visual talk.

These birds, small members of the weaver family, are infamous for their numbers. They're the world's most abundant wild bird, with a billion and a half individuals in the breeding season. Massive flocks of queleas darken the sky the way passenger pigeons once blotted the sun in North America, one of nature's fantastical spectacles. But also, so hugely destructive to agricultural crops such as millet that the queleas are known locally as "Africa's feathered locusts."

Less well known is the dramatic variability of their facial plumage and how they use it to communicate—to broadcast their identity and keep peace with their neighbors. Breeding males have a bright red bill enclosed by a face mask that varies from white to black, and in size from nothing at all to a very broad stripe. Surrounding the mask are colored feathers ranging in hue from red to yellow, and in size from a thumbnail patch to a huge swath of breast and belly. The combinations of pattern and color are almost endless.

The variation in quelea facial plumage is extreme by any measure, says James Dale. Scientists usually chalk up disparities in plumage color to differences in physical condition. Bright, colorful plumage is a flag of fitness, an honest indicator of quality. It's hard work to keep feathers bright with rare pigments from the environment. Birds in better shape flaunt more brilliant hues.

When Dale began studying the bird years ago for his PhD thesis, he

expected the rainbow of quelea faces would reveal a neat correlation between color and fitness. Brighter face colors, healthier bird, with a better
chance of mating and reproducing. For years he struggled to find a link.
It was not easy work. "The birds nest in these horrific thorn trees," he
told me, "one of those 'wait-a-minute' (while you disentangle yourself or
you'll shred your clothes) trees." No matter how he diced the data, Dale
could find no correlation at all between fitness and facial plumage and
almost gave up on the project. He stuck with it, though, and discovered
something even more interesting: The differences in quelea facial plumage are signatures of individuality for the birds, a kind of name tag that
announces their identity in a swarm of strangers.

Red-billed queleas nest in crowded colonies of millions of birds because there's safety in numbers. "In such a massive flock, chances are
your chicks won't be eaten," says Dale—though it's still a sort of grim
lottery. Eagles haunt the breeding grounds "and will tear open a nest,
eating one nestling after another like a can of cherries." The birds don't
try to defend their nests from predators. They just breed very quickly
and simultaneously, right after a heavy rainfall when the annual grasses
that provide their seeds grow lushly. The males arrive at a breeding area
at night, and by the next morning, the thorn trees are packed with a huge
congress of queleas building their nests. A single tree may hold hundreds
of nests. "They'll start in this explosive manner, all frantically weaving
their nests at the same time," says Dale, building en masse for three days.
"And it's just this absolute noisy chaos and havoc, and then by the eighth
day, all the nests are full of eggs."

The thing about nesting in such crowded conditions—up to five million nests within a couple of kilometers—is that there's a real danger that
a random male will usurp your nest. "These are really cheeky, aggressive
birds, so they're constantly testing one another, stealing grass and stuff,"
says Dale. Moreover, everything happens so fast and so synchronously,
with every bird doing the same thing at the same time, that there's no
opportunity to establish boundaries and territory. "They don't have time
to sit there singing all day the way North American songbirds do," says

Dale. So the quelea form little neighborhoods surrounded by familiar birds, individuals they know will stick to their own nest and not try to steal from them. A newly arrived, unfamiliar male poses a bigger threat than a familiar neighbor going about his own nest building. It's called the "dear enemy effect." The birds get to know their neighbors very quickly. One quelea signals its identity to another by flashing its face, making it easy for the neighbor to quickly learn who it is. "Once that's settled," says Dale, "everyone can stop harassing each another and get down to the business of nest building."

As it turns out, it's the quelea's bright red bill that signals quality, embedded in their colorful feathers of identity. Imagine all that information, all that messaging, contained in a bird's face. But really, should we be surprised? Recognizing individuals is the foundation for social relationships, and queleas are highly social creatures.

Birds may prattle and rant with feathers and body poses, but far and away the most common, the most extreme, and the most complicated kind of bird babble is vocal.

In the dark hours before dawn, the Pilliga woodlands of inland New South Wales buzz, screech, whistle, ping, trill, and *bark* with birdsong. This is the largest surviving remnant of native forest on the western slope of the Great Dividing Range, and the warm-temperate woods of ironbark, red gum, and Pilliga box are bird rich with grey-crowned babblers, laughing kookaburras, white-winged trillers, rufous whistlers, even barking owls—birds with names that evince their songs and calls. It's a startling racket of bird sound, but not unusual in Australia. That some of the bird world's strangest, loudest, most extraordinary voices come from the great southern continent is no accident. As Tim Low will tell you, Australia is where birdsong began. DNA analysis has revealed that songbirds, as well as parrots and pigeons, evolved on the continent and radiated outward, spreading around the globe in successive waves. The birds that carol the dawn chorus at my home halfway around the world in central Virginia—American robins, mockingbirds, warblers, sparrows, cardinals, finches—all descend from the early passerines of Australia, and like Pilliga's birds, they all talk at once.

The dawn chorus has always seemed to me a baffling behavior, everyone singing at the same time, louder and more energetically than at any other time of day, like a poetry slam where everyone simultaneously yells out their offerings. The chorus begins as early as four a.m. and lasts several hours until the sun rises and temperatures warm. It often begins with larger birds, such as doves, thrushes, and robins in temperate zones, and in Australia, the big Australian magpies, butcherbirds, and kookaburras. But more important than body mass is eye size. Scientists in the UK found that birds with larger eyes and greater visual capability at low light sing earlier than others—true in neotropical habitats, too. Karl Berg studied the dawn chorus in a tropical forest in Manabi, Ecuador, and learned that foraging height and eye size were the best predictors of time of first song, with big-eyed canopy-foraging species singing earlier than smaller-eyed ground-foraging species.

Why birds sing so intensely before dawn is not well understood. It may have something to do with the advantages of acoustic transmission in those dark early hours. Cooler temperatures, calmer air, less ambient noise from insects (and traffic), allow a bird's song to travel farther, the better to stake its territory—at least in northern species—or to broadcast its presence to prospective mates. Or maybe it's that predators are less of a threat. Or maybe the birds are up anyway, and the low light makes foraging difficult, the still air not suitable for migrating, and insects not yet out and about. So why not sing? Maybe the birds are practicing, warming up for the day. Or maybe it's just their way of announcing, "I survived the night!"

Andrew Skeoch, an Australian wildlife sound recordist, views the dawn chorus as a communal and collective phenomenon in which individual birds negotiate and affirm their relationships while minimizing conflict. "It's a reaffirmation of place and belonging every morning with mates, family groups, neighbors, and flocks," he says. "By avoiding physical confrontations, the dawn chorus reduces risks and stress and conserves energy. It's a tapestry of vocal behaviors," he says, "and it may be the greatest evolutionary achievement of songbirds, allowing them to coexist and to become the wildly successful and diverse group they are."

Bird songs and calls range from the odd comical cluck and rattle of the willow ptarmigan and the soft piping voice of the pardalote, barely audible, like whispered gossip, to the elfin chucklings of Leach's storm petrels, the gongs of three-wattled and white bellbirds, the loud trumpeting of the southern screamer, the organ-like caroling of Australian magpies, and the gorgeous, haunting nocturnal solo of the pied butcherbird, which may go on for seven hours. Butcherbirds are the Sweeney Todds of the bird world. They do dastardly deeds—skewering small birds and other animals for dinner—but they sing like seraphim, sometimes in trios. So spectacular and haunting is this bird's song that violinist and composer Hollis Taylor worked for a decade recording it and transforming it into music. In 2017, the striking composition she created incorporating her field recordings, "Taking Flight," was performed by the Adelaide Symphony Orchestra.

Among the weirdest calls I've ever heard comes from the green catbird, a handsome little bird with such perfect camouflage, a mottled green and fawnish brown, that it's more often heard than seen in its rainforest home. Its call sounds like a cross between a yowling cat and a wailing toddler. The first time I heard it, I thought, "What in the world is happening to that poor child?"

Science is just beginning to parse the complexity and meaning of bird vocalizations. Even common species such as American robins make more than twenty different types of sounds, most of which remain mysterious in purpose. The simple honk of a goose, it turns out, contains unexpected richness and intricacy; and calls that sound simple and uniform, such as those of penguins, vary in their acoustics, helping penguins recognize one another and choose mates.

The vocalizations of most songbird species differ from place to place, forming local "dialects" just like human accents, distinct and long-lasting regional and cultural differences in the structure and composition of songs. These dialects play a role in courtship—females of some species prefer males with songs that include syllables from their own song vocabulary—and also, in resolving territorial disputes, allowing birds to distinguish between local and foreign individuals and settle conflicts without fighting. The ornithologist Luis Baptista was among the first to recognize bird dialects in his studies of the white-crowned sparrow in coastal California. Called the "Henry Higgins of the bird world," Baptista could pinpoint the geographic origin of a sparrow and its parents just by listening to its song. So localized were the accents of these birds, he said, that one could stand facing the Pacific and hear the songs of one dialect with the left ear and a different one with the right.

The voice box of birds is a structure called a syrinx, buried deep in a bird's chest cavity. Sound emerges when the membranes of the syrinx vibrate, shifting the flow of air through the organ. The syrinx in birds varies from the bulbous resonance chambers and long looping trachea of ducks, geese, and swans—up to twenty times the expected length—which produce sound that exaggerates their body size, to the tiny pair of chambers in songbirds, controlled by delicate syringeal muscles. Some songbirds have such fine control over the multiple muscles in both sides of their syrinx that they can produce different sounds at the same time, in essence, singing a duet with themselves. This accounts for the rich caroling of the Australian magpie and the glorious fluted warbling of the wood thrush.

It was once thought that the hearing of birds was limited to a smaller frequency range than human hearing. Only lately have we learned that some birds such as the vinous-throated parrotbill and the black Jacobin hummingbird make sound in the ultrasound range, beyond human hearing, suggesting they may also be able to perceive sounds "invisible" to our ears. Birds are generally better at recognizing sound than we imagined, keenly sensitive to variations in pitch, tone, and rhythm in the sounds of their own species, which allows them to identify fellow birds not just as members of their own species, but as individuals within their flocks, even in noisy, chaotic conditions.

A fine example of birds using sounds to recognize individual members of their own flocks: the contact calls of budgerigars, which vary subtly from one bird to the next. Like queleas, budgies live in huge flocks. There was a time, in the 1950s and 1960s, when the birds were described as "so thick on power lines that the wires sagged nearly to the ground with their weight." Their contact calls allow them to identify their mates and their flock. They can continue changing these calls as adults, modifying them to match their mates or other flock members as the birds move from one social group to another.

Budgies and other birds learn their songs and calls through a process very similar to the way we learn to speak. It's a process of imitating and practicing called vocal learning, and it's extremely rare in the animal world. Vocal learning in birds begins early, just as it does in humans. By the last trimester of pregnancy, a human fetus can memorize what it hears from the external world and is especially sensitive to melody in both music and language. This appears to be true for some birds, too. The embryos of certain species can hear through the shell of the egg; in response to the voice of a parent, their heart rate increases. As a defense against brood parasites, superb fairy-wrens learn special vocal passwords from their parents when they're still in the egg. Scientists have found that at least five days before they hatch, the unborn fairy-wren chicks learn to imitate the call. Zebra finch parents can tell their young while they're still developing in the egg that it's hot outside. This is vital information for a growing chick. In hot climates, birds need to be able to lose heat, which is easier with a smaller body. When zebra finch parents are breeding in a hot climate, and the nest hits a temperature above 80 degrees Fahrenheit, they'll chirp the news to their unborn chicks in the last third of the incubation period—the moment when the embryos are developing their temperature-regulation system. In response to these "hot calls," the chicks will actually curtail their growth and emerge smaller—an adaptive advantage in the heat.

Birds cry like children, grunt like pigs, meow like cats, and sing like divas. They speak in dialects and carol in pairs and choruses. They glean all sorts of information from calls and songs—a singer's species identity, its geographic origin, group membership, even its individual identity. And they use sound in ingenious ways—to share information, negotiate boundaries, and influence one another's behavior.

Bell miners use their calls to create literal walls of sound that keep out other species. At the Gap Creek Reserve in southeastern Australia, the woods are full of the tinkling and twittering of scarlet honeyeaters and other small forest birds and the plaintive *crees* of silvereyes. But walk a little way down the Bellbird Trail through the eucalypts, and that forest chatter is replaced by a single sound, a bell-like *ting-ting-ting-ting*. It's the classic call of the bell miner, a pugnacious honeyeater with a dark streak running downward from its bill, like a little frown. Once you enter this bird's fiefdom, the avian chorus is diminished to a single chime that sounds everywhere from high in the canopy.

One bell miner is a delight to hear. And for a minute or two, a colony of forty pinging away is wondrous, as if stars could talk. But then the chorus starts to irritate, like ringing in the ears or the *plink, plink, plink* of a dripping faucet. Unlike North American birds, where territorial calling is seasonal, these birds start pinging at dawn and go right through to dusk every day of the year. "It's one of the world's most constant, pervasive animal sounds," says Tim Low. "The miners are saying, 'Stay out; if you come into the colony, you'll get attacked.'"

Bell miners are aggressive defenders of their territory, known for attacking larger species such as kookaburras and pied currawongs and for completely banishing small species. Birds such as fairy-wrens and scrubwrens that forage low in the understory, the miners will tolerate. But birds that share similar areas of the forest and similar foods, like pardalotes, are held at bay by the sound. Just beyond the boundaries of their ringing, pinging turf live normal assemblages of birds. The miners are able to suppress the numbers of competing species in their territory, which they hold for years at a time.

One small bird that coexists with the bell miner by keeping a low profile has its own notable vocal talent, used for an entirely different purpose. The eastern whipbird, a slender olive-colored bird with a white cheek patch and a cute little black crest, throws its whole body behind its voice. And what a voice. The whipbird's dramatic call is considered such a signature sound of the rainforest that it's used in movies for jungle scenes. The distinctive and enchanting two-note "whipcrack" call, a kind of thin whistled *whoooopp* capped by a sudden, loud *chew, chew!* is actually a duet, male and female singing in a call-and-response so precisely timed and seamlessly delivered it sounds like one bird calling. The male initiates the call, and the female responds within a fraction of a millisecond.

Why would a bird bother with such a highly choreographed saraband of sound?

There's a female-biased sex ratio in this population—fewer male than female whipbirds—so there's actually competition between females for males, explains Naomi Langmore, who studies bird behavior at Australian National University in Canberra. A female whipbird may use the duet to defend her exclusive position in a partnership with the male. It's called mate guarding, and it's an unusual strategy for female birds. But it makes sense for a female to use song to claim "ownership." "Every time a male sings, the female has to sing, too, to say, 'Hey, but he's got a mate,' so that other females don't come in and try to steal him," says Langmore. "And conversely, when the male answers the female song, he could be saying, 'Hey she's got a mate, so don't try to steal her away.' For a long time, it was thought the duet was produced by the male alone, so perfect is their timing."

Other species may duet for reasons more like the bell miner's. "There are species where the main function is territorial defense," Langmore explains. "By coordinating really nicely, they're saying, 'Look we're a formidable team, a really established pair, we can do this beautifully timed, coordinated duet, so we've got a long history together on this territory, and you're not going to intrude on it.'"

Whatever its purpose, duetting in birds is a marvel. It occurs in about 16 percent of species—mainly in the tropics—and that percentage is distributed across almost half of bird families, so the practice seems to have originated multiple times. It ranges from the relatively simple, overlapping songs of stripe-headed sparrows to the intricate, tightly coordinated duets of many neotropical wren species, where male and female sing alternating song phrases, each member of a pair fine-tuning the timing and type of phrases it sings to match the phrases sung by its partner. These more precisely coordinated, complex duets are the closest analogue we have in the animal world to the structure of human conversation.

A typical human chat is a study in seamless give-and-take. Unwritten rules dictate that one individual speaks at a time, with little or no gap of silence and, ideally, little overlap. Think how strange it sounds when these rules are broken—say, in a radio interview, when there's a lag time between a question posed by an interviewer and the subject's delayed response. Long pauses between turn taking are uncomfortable. In conversation in nearly every human language, each turn lasts for around two seconds, and the typical gap between them is just two hundred milliseconds.

In the duets of canebrake wrens, the timing of male and female give-and-take is even more precise. Karla Rivera-Cáceres studies the highly coordinated duets of canebrake wrens in the forests of Costa Rica. So impeccable is their timing that canebrakes answer their mates within sixty milliseconds, around a quarter of the time it takes for a human to chat back, and with even less overlap of phrases, 2 to 7 percent, versus our 17 percent. To the uneducated ear, the multiple and varied to-and-fro phrases of a canebrake pair sound like a single bird.

Singing this kind of perfectly coordinated, delicately timed antiphonal duet presents extreme perceptual and cognitive challenges. The duets are composed of three different categories of phrases, specific to each sex. Male and female both possess a repertoire of up to twenty-five types of song phrases in each phrase category, which can be combined and repeated to form a song. Like some other duetting birds, each pair follows a strict duet "code" that's specific to that pair, stipulating which song phrases answer each other. Moreover, to coordinate their duets, as Rivera-Cáceres discovered, the birds modify their singing tempo based on the phrase types that their partners are singing. Coordination requires both birds to pick the appropriate phrase and time it exactly right, all in a matter of milliseconds.

Happy wrens can achieve this feat in darkness, without any cues except sound. Christopher Templeton of Pacific University and his colleagues captured male happy wrens in the dry forests of western Mexico and held them in a cage in a lab overnight. Then, first thing in the morning, the team played different songs from the repertoires of the birds' mates. "Miraculously, the male birds in the dark of their cages were able to respond to the playbacks of the very first song phrase," says Templeton, "correctly recognizing the song phrase type sung by their mate and responding with the correct reply—all within half a second."

Rivera-Cáceres found that the duet codes of adult male and female canebrake wrens are flexible and can change when the birds have to find a new mate. For birds entering a new pairing, there's a steep learning curve, but eventually, the two birds create one seamless song.

In 2019, scientists working with birds in the wild discovered that when two birds sing these sorts of precisely coordinated duets, their brains actually synchronize. A team of researchers from the Max Planck Institute of Ornithology fitted pairs of white-browed sparrow weavers (a species native to eastern and southern Africa) with little backpacks carrying miniature microphone transmitters that recorded both vocalizations and neuronal activity during their duets, then released them into their natural habitat and recorded them singing hundreds of duets while perched in trees. The team discovered that in both partners, the nerve cells in the vocal control areas of their brains fire in time together so their two brains essentially function as one.

This ability to engage in such exquisitely precise pitter-patter is not automatic or innate. It's learned, just as it is for humans. We first get a feel for the exquisitely timed turn taking of conversational rhythm when we're still babbling babies. And it happens, the same is true for the canebrake wrens when they're learning from their parents how to duet. Young birds overlap more, trampling on the lines of their parents, but they get better over time. And Rivera-Cáceres has found that young wrens copy the phrase types used by their same-sex parent to answer each phrase type of the opposite-sex parent. So there's cultural learning going on as well.

I t was duetting that first gave scientists an inkling how they might have gotten it wrong on female birdsong.

For centuries, it was thought that male songbirds alone used song as they used fancy plumage and elaborate tails, to attract females and compete with rival males. The female role was to listen and choose and, in so doing, exert influence through sexual selection, driving the evolution of elaborate singing by preferring males with ever more melodious and complex songs. This was considered a classic example of the power of sexual selection to generate extreme sex differences in brain and behavior. Instances of female song were largely dismissed as atypical—rare exceptions or the outcome of hormonal abnormalities.

All that changed when the international team of researchers led by Karan Odom surveyed the occurrence of female song in 1,141 species from all over the world. The team suspected that male-only song was not the whole story by any stretch. Most songbird species live in tropical regions, which have been less studied. And female song is more common in

Australia and surrounding areas, where songbirds originated. There, it's a much more egalitarian setup. Sure enough, the results of the study, published in 2014, showed that female song occurs in more than two-thirds of surveyed songbird species and families—and that it's structurally similar to male songs (long and complex) and used for similar purposes. In the songs of male and female superb starlings of eastern Africa, for instance, there's no difference in the structure or the number of motifs that each sex sings. Both sexes in this highly social species sing all year long and may do so to signal identity and establish rank within their groups. The study made it clear: Birdsong is not just a male thing. And elaborate song did not evolve just through sexual selection—females choosing males for their vocal prowess—but through the broader process of social selection: both sexes competing for food, nest sites, territories, and mates. Most important, the scientists showed that female song was probably common in the ancestors of most songbird species, even the ones in which females today sing less or not at all. In other words, it's not that these females never sang; it's that over evolutionary time, they've lost their song.

Why?

Here's where duetting comes in. The temperate regions such as North America and Europe, where duetting is rare, and female song in general is less common, are the regions that host a lot of migrants. "It's the migratory birds that have lost female song," explains Langmore, who was part of the study. "Migrants have a very different pattern of territoriality and pairing than in the tropics. Typically, the male will arrive on the breeding grounds singing his head off, and the females will fly in and listen, and they'll plunk down on the chosen male's territory. Then they'll have a very short breeding season. They just go for it like crazy, and then they leave."

Resident birds, on the other hand, like those in the tropics, have to defend their territory all year long. If, over the years, a partner dies, the remaining bird has to be able to defend that territory and attract a new mate. "And there can be divorce, as well," says Langmore. "Whichever bird is left has to be able to sing to defend its territory or attract a new mate. So it seems like it's quite a recent thing that females have lost song in these migratory groups."

It's this sort of news that turns the room ninety degrees. Maybe the question is not why some female birds sing, but why some don't—or whether in fact they do sing, and we just haven't been listening.

Chapter Two

CAUSE FOR ALARM

The Australian National Botanic Gardens seems an idyllic place, a big swath of bushland near the center of Canberra. Wide tracks wander through more than 6,500 plant species representing the native habitats of Australia—eucalypt forests, spinifex grasslands and saltbush scrub of the continent's red center, even a rainforest gully lush with plants from the diverse rainforests that extend along the country's east coast from Tasmania to Queensland. Here are kangaroos and swamp wallabies, bats, brush-tailed possums, and sugar gliders. The main path is lined with bright red flowering waratahs and grevilleas, and the yellow spiked flowers of banksia. These are manna for the many nectar-feeding birds that frequent the garden—honeyeaters busily darting from flower to flower, acrobatic eastern spinebills that use long, elegant bills to probe the flowers, and raucous red wattlebirds, named for their somewhat less elegant red wattles, or lobes of skin on the sides of their necks. Little herds of white-winged choughs scratch among the mulch beds. In the rainforest gully, crimson rosellas forage on fruits or nibble on the spore capsules beneath tree ferns, and on the Eucalypt Lawn, beneath a phalanx of blue gums, brittle gums, blue mallees, ironbarks, and gympie messmates, are twitterings of superb fairy-wrens.

Noticeably absent from the gardens are bell miners and noisy miners, which is why the place is swarming with small birds usually kept at bay by the aggressive miners, among them, the tiny spotted pardalote, with its soft piping voice drifting down from treetops, where it's picking insects from the leaves of eucalypts.

It all appears very bucolic.

But linger here for a bit, and you sense trouble. A harsh, high-pitched rattling alarm call erupts from the eucalypts, loud and sudden, and ricochets around the gardens. To the birds here, it's the sound of danger.

There are snakes in this garden, literally and figuratively: deadly eastern brown snakes that will snatch the eggs and nestlings from a wren's nest. Fan-tailed cuckoos, which parasitize the nests of the little birds here and evict their young. Introduced mammals like feral cats and European red foxes. And, worst of all, birds of prey, kookaburras, currawongs, brown goshawks, and the deeply feared collared sparrowhawk, a bird that specializes in catching small birds in flight. It's known for its big, bright piercing yellow eyes, its relentless tail chase, and its very long middle toe used to clutch prey before it's killed, plucked, and eaten.

The strident alarm call belongs to the New Holland honeyeater, a handsome black-and-white bird with yellow wing patches, modest in size compared with the outrageously loud pitch of its voice. Its call is a kind of early-warning system of impending danger. The alarm spreads from honeyeater to honeyeater, tree to shrub to tree, across the gardens in a fast-traveling wave of raucous real-time alerts that tracks the progress of a sparrowhawk.

"New Holland honeyeaters are constantly monitoring the gardens," says Jessica McLachlan, who studied the birds for her PhD thesis. The instant a honeyeater spots the sleek profile of a sparrowhawk flying in low and fast over grevilleas and gum trees, scanning for unsuspecting prey, it unleashes a torrent of warning calls, alerting other birds to the presence of the hawk—fairy-wrens and scrubwrens, brown and striated thornbills, pardalotes, and other honeyeaters, which issue their own alarm calls, until the woods and shrubs are ringing with alarm. Everyone knows a predator is present, and because of this communication network among its prey, there's a good chance the hawk will go hungry.

McLachlan has found that the honeyeater's alarm calls are not in fact simple cries of warning, but a complex language rich with meaning. "These birds really go extreme, with up to ninety-six elements in a single call," she says. *Ninety-six?* "That's the most we've heard; the median is much less than that. But these birds are saying a lot more than we ever suspected."

McLachlan has discovered that the honeyeater's calls encode highly specific information. Moreover, the birds that are listening are capable of decoding and understanding these complex messages. It's a super-sophisticated signaling system, and McLachlan has found an ingenious way to unravel its mysteries.

McLachlan was a self-described nature nerd from the start. She grew up in South Africa, in what she describes as a natural paradise, "with crabs in the curtains, snakes in the fishpond, scorpions in the tree bark." There were sunbirds in the flowering erica shrubs, a Knysna warbler that "couldn't distinguish between morning and midnight," and sombre greenbuls that, she says, were always the last bird out of bed.

McLachlan knew the greenbul's morning habits from research she conducted while still in high school. For a science fair project, she and a friend looked at how the size of a bird's eye affects how early it sings in the dawn chorus. The paper by scientists in the UK had come out suggesting a correlation, and the two students decided to look at South African birds. "The theory was that the bigger your eye, the more light is let in, the brighter the world seems and the earlier you sing," says McLachlan. Despite the "very annoying sombre greenbul—which was our massive outlier and despite its big eyes always started last in every chorus"—the pair did find the expected correlation.

"That was pretty exciting," she says. But the process—two fourteen-year-old teenagers having to get out of bed before dawn nearly every morning—was not. "Both of us swore we would never do anything on birds ever again." However, the very next year, McLachlan came up with another project on bird vocalizations. She ended up studying biology at the University of Cambridge, and when it came to picking a topic for her PhD, New Holland honeyeaters popped out to her. "They're really extreme in their vocalizations, so I thought they'd be a cool system to look at."

McLachlan calls the honeyeaters her "little informants." "I hike a lot, and when I'm in Australia, if there's ever anything interesting to see, any raptors, the honeyeaters will let me know," she says. "I rely so much on them to tell me what is going on out there that when I'm in another country, the UK, for instance, I feel like I've actually lost a sense."

Birds rely on the honeyeaters, too, as a matter of life and death.

"The world of birds is a dangerous place," says Rob Magrath, who studies behavioral ecology at Australian National University and oversees research projects by numerous PhD students, including McLachlan. "Life as a small bird is like being a human in Jurassic Park," he says. "You're constantly on edge, constantly wary of what can eat you, including hawks ten or twenty times your size."

No wonder birds have evolved numerous strategies for raising alarm. Magrath and his students have found that some birds signal danger with just their wings. Crested pigeons have specially modified wing feathers that produce distinct notes during escape from predators, prompting other birds to flee. Sometimes the warning isn't a sound at all but the silence between sounds. When noisy, social birds like red-winged blackbirds suddenly get quiet, it's a strong signal there's danger nearby. But most birds call out vocal warnings. We've all heard them: the quick, repetitive crow-like call of a blue jay, the loud metallic *chit, chit* of a northern cardinal, the sharp little *shree* or *yeep* of an American robin, and the *tsee* of a great tit.

For decades, research looking into alarm communication in birds has consisted of doing playback studies, a powerful experimental approach to studying bird vocalizations and their functions. Scientists play recorded bird vocalizations through loudspeakers and then observe a bird's responses, whether or not it flees.

McLachlan added video recordings to her study, which allow her to view extremely fine-scale responses in small birds that are really difficult to see.

"I thought I could get a lot more information if I filmed the birds," says McLachlan, so she devised a "less than elegant" system, in her words, for mounting an enormous amount of equipment on her body—microphone attached to her shoulder, binoculars around her neck, amplified speaker playback device strapped to her waist, and—at first—a video camera duct-taped to her hat, with one eye completely covered by the screen. "I had to make sure the screen was right, and the video was pointing at the birds so I could look at exactly what they were doing and spot the timing of their responses." Later, she moved to a backpack device.

When she goes about her work in the botanic gardens, wearing her full array of equipment, she is sometimes mistaken for an exhibit or a piece of "art in place." To get good footage of the birds, she has to stand stock-still despite aching shoulders, stiff neck, and gawking onlookers. Visitors strolling along the paths between the habitat displays remark on the strange lifelike sculpture of a tall, thin figure heavily adorned with electronic equipment standing perfectly still near the flowering shrubs of grevillea and banksia. McLachlan says she comes off looking like an over-the-top 1960s model of a futuristic scientist or tourist. Most people are pretty sure the figure is meant to represent a male scientist, especially in winter when she wears a ski mask to keep her face warm, showing only her mouth and eyes.

"This is a very strange exhibit they have up in the gardens!" people say.

"What is the figure supposed to represent?"

"Oh my god, it's blinking!"

Children come over and poke her to see if "it" is alive.

At one point the garden had a dinosaur exhibit on display, and people thought McLachlan was a model of a time-traveling explorer. In winter when she was dressed in black, covered in electronic gear, "she looked like a member of a swat team," Rob Magrath recalls, "and on at least one occasion, caused terror when she walked into a toilet block!"

The hassle is worth it. All that equipment, all that time frozen in one position, has given McLachlan an unprecedented view of some truly remarkable behavior.

It's generally thought that birds have two kinds of alarm calls: mobbing calls and flee alarm calls.

One summer a pair of northern cardinals built a nest in the crape myrtle shrub right outside my office window. I remember one breezy morning in mid-July. The female had been brooding three eggs for several days. She was off the nest when I heard a racket of birds fluttering, churring, spishing, scolding around the crape myrtle. I looked out and saw chickadees, tufted titmice, cardinals, robins. The whole avian neighborhood was suddenly up in arms with loud *chit*s and *yeep*s and *shree*s. It was deafening. I looked around. On the rooftop next to the nest sat a crow. I looked in the nest. There were only two eggs.

These were mobbing calls—abrupt, short, loud, and repetitive—alarm calls made in response to predators that are not moving at high speed and so are not an immediate or intense threat—usually a terrestrial predator like a snake or cat or, in this case, a perching bird. The call alerts other birds and signals them to fly toward the source of the call and join in with their own mobbing calls, or attack or mob the predator to drive it away. "There's a predator here! Come help me harass it!"

High-pitched flee, or aerial, alarm calls, on the other hand, usually mean there's a predator in flight, which is a lot more dangerous for a bird. These calls are typically in a narrow bandwidth, with a lot of up-and-down amplitude, making the sounds harder to locate, especially for raptors with relatively poor hearing in that frequency range. Small birds use flee alarm calls to alert other birds to imminent danger from above, signaling them to freeze or take immediate cover, while not boosting their own chances of being snatched by a predator.

Flee alarm calls send birds away from a threat; mobbing calls bring them toward it.

It's worth pausing here to consider the paradoxical nature of mobbing. Small birds flying *toward* a roosting owl or a cat, even throwing their bodies against it?

It seems counterintuitive—time consuming, energy expensive, and downright dangerous.

Crows are among the most frequent mobbers, swooping and dashing down on a hawk from above and behind it, always keeping the menace in sight. Gulls often resort to the practice, too, with an unusual twist: vomiting on the predator with keen aim. Colonies of fieldfares fire from another orifice, ejecting feces on a predator in such volume and with such accuracy that the threatening creature is literally grounded or stopped in its tracks. If enough of these droppings-bombs hit their target, they can soak a bird's wings so it can't fly.

In a paper entitled "One deleterious effect of mobbing in the Southern Lapwing," J. P. Myers suggests the risks involved. He reports seeing a pair of lapwings tending chicks that were foraging by the water's edge in Argentina. When a crested caracara flew over them, both adult lapwings mobbed the caracara, dive-bombing it from above. During one of the lapwings' dives, the caracara flipped over in flight, stuck his legs out, grabbed the lapwing in the air, and flew off with it over the pampas.

Why mob if it's so dangerous? It's a good way to expose a predator and to drive it off, especially when young are at risk. It's also a way to impress on inexperienced birds a predator's dangerous nature, offering so-called teacher benefits. A naïve bird watching other birds mob a threat teaches that bird to fear it, too, and to either avoid it or mob it more strongly, creating more knowledgeable informants and mobbers. There's strength in numbers.

The classic experiment demonstrating that birds can learn about threats from fellow birds was conducted by German zoologist Eberhard Curio. Curio showed that European blackbirds learned to regard a harmless bird as a predator by observing other birds mobbing it. He put one "teacher" blackbird in a box and showed it a species that was a real threat—a model of a little owl—which evoked a powerful mobbing response in the teacher bird. At the same moment, he showed another blackbird—the "observer," or student—a model of a bird that's harmless to blackbirds, a noisy friarbird. The student bird could see both the teacher and the friarbird but not the little owl. Almost instantly, it learned to fear, and mob, the friarbird. Remarkably, Curio showed that this recognition and fear of an enemy could be culturally transmitted from bird to bird, along a chain of up to six individuals. Bird number six in the string mobbed the friarbird with a ferocity equal to the original teacher bird—a superb example of social learning.

Given that birds need to respond to different threats in different ways, mobbing or fleeing, it makes sense that they would evolve different types of alarm calls. But communicating in this way—describing the specifics of a predator, whether it's arriving by air or by ground—is called functionally referential signaling, and it was considered a big deal when it was first discovered in birds. For most of the past century, scientists thought the ability to refer to a specific object or event in the environment was unique to human communication. Animal signals reflected only an animal's "internal state."

That changed in the late 1970s when scientists working with African vervet monkeys at the Rockefeller University found that the vervets produced distinctive calls for different predators—leopards, martial eagles, pythons, and baboons—and responded appropriately with different behaviors, climbing a tree for a leopard alarm call, for instance, or scanning the sky for an eagle alarm. A low-pitched, grunting *rraup* call signaled a swooping eagle and prompted the monkeys to look up and flee to a bush for cover. In response to a snake *chutter* call, they stood on their hind legs and looked down, searching for a python.

A decade later, this ability to designate different kinds of danger was first demonstrated in domestic chickens, which produce a high-pitched screech in response to flying raptors and a kind of deep throaty garble when they see ground predators such as raccoons.

"This categorization of threats into those that are flying and those that are on the ground seems to be a pretty common strategy among birds," says McLachlan. The smooth-billed ani of the Caribbean islands gives *chlurp* calls in response to flying birds of prey and *ahnee* alarms in response to terrestrial threats like cats and rats. Japanese tits, relatives of chickadees, use two different alarm calls to specify the type of predator attacking their cavity nests. When a jungle crow approaches the nest to pluck nestlings from the entrance with its beak, the tits make a *chicka* call, which prompts the nestlings to crouch down inside the nest cavity. If a Japanese rat snake wriggles up to invade the nest, the tit's *jar* call makes the chicks jump out of the nest to escape.

When Toshitaka Suzuki, a researcher at Kyoto University, discovered this, it inspired him to explore whether this sort of specific alarm call might actually cause the listening bird to summon a specific image in its brain, in this case, of a snake. In human speech, words referring to a specific object usually evoke a visual mental image. Think *moon* or *dog* or *snake*. Suzuki found evidence that Japanese tit alarm calls create just this kind of search image in the heads of listening birds. The tit's *jar* call causes the birds to search for snakes and to become more visually responsive to any object resembling a snake. It was the first evidence that a nonhuman animal could visualize something referred to in a vocalization.

A larm calls may also encode what a predator is *doing*. This was first discovered in Siberian jays, social birds that live in family groups in the boreal forests of northern Eurasia. Predators are a big problem for these birds. During their first winter of life, more than a third are taken by owls, pine martens, sparrowhawks, and especially goshawks, which account for 70 percent of all kills. Goshawks hunt in three phases—perching in a tree and scanning for prey, making search flights to the next perch, where they scan some more, and then, finally, bulleting through the brush to take their prey by surprise. The jays have evolved a sophisticated system of three different alarm calls specifying the hunting phase of the hawk—whether it's perched, searching for prey, or attacking.

McLachlan has found this is true for New Holland honeyeaters, too. "With eyes shut, you can basically track what a butcherbird is doing— whether it's perched or in flight—just by listening to the change in their alarm calls." Now we know that other species also use a variety of calls to spread the word about a predator's behavior. Using a "hawk glider," an aerodynamic foam model of a hawk, Rob Magrath and his colleagues found that noisy miners produce one alarm call in response to a raptor when it's perched (and poses little immediate threat), and a second one in response to the very same species when it's flying. "If you throw out a hawk model, the miners will go straight to the aerial alarm calls," says Magrath. "When it lands on the ground, they switch instantly to the 'churr' mobbing calls. And similarly, if you plunk a predator on the ground, they'll give the mobbing calls, and as soon as the thing leaves your hand when you start to throw it, they'll switch to the aerial call. So there's very clear encoding of predator behavior."

The specificity goes even further. In white-browed scrubwrens and fairy-wrens, the number of notes encode information about how far away a predator is. The *chickadee-dee-dee* mobbing alarm calls of black-capped chickadees contain messages—coded in the number of *dee*s at the end of the call—about the *size* of a predator and hence, the degree of threat it represents. More *dee*s means a smaller, more dangerous predator. A great horned owl, too big and clumsy to pose much of a risk to the tiny chickadee, elicits only a few *dee*s, while a small, agile bird of prey such as a merlin or a northern pygmy owl may draw a long string of up to twelve *dee*s.

When I first learned this about the chickadee, it changed the way I heard all birds. What I'd taken as random chirps were in fact sophisticated signals to other birds, tweets of intelligence.

More notes means a greater menace. This rule holds true across species and calling contexts and even continents. It makes sense. With a long string of notes you get your message across and also reduce the risk of false alarms. Think about what would happen with the opposite system: If a one-note call meant big danger, birds would flee at the start of what could well be a longer, less urgent call, resulting in a lot of wasted energy. "But New Holland honeyeaters really go all out with the number of notes in their aerial alarm calls," says McLachlan, packing them with multiple notes. I thought of the extreme example of ninety-six. If a small bird sat through even a fraction of those notes to get the "danger" message, wouldn't it be toast?

Yes, says McLachlan, in this way, it's counterintuitive. "Hunting hawks move really, really fast through the environment, at speeds up to eighty feet per second," she says. "You need to make really quick decisions, so it seems incredibly weird that a bird would sit there listening and thinking, 'Hmmm one note, not so bad; two, three, four, five, six . . . oh crap! It's upon me now, and I'm going to die!'" Even a split-second delay in response can result in considerable ground gained for the hunter. "In this context it's really puzzling that birds would use more elements to indicate that a predator is closer," she says. "More elements take longer to produce—and longer to receive, to get the message and realize how grave the danger is."

How do birds resolve the dilemma of communicating reliably about the degree of danger but also rapidly enough to allow the listeners to escape a dive-bombing hawk?

McLachlan is exploring the question in New Holland honeyeaters. Here's where the video recording equipment comes into play. "In addition to looking at what the responses were, I could look at how rapidly they were occurring," says McLachlan. "When I watch the videos, it blows my mind how insanely fast they respond to these calls—literally faster than the blink of an eye." She's now taking a closer look at the first element of the alarm calls. Is there variation in its acoustic structure that the birds might use to accurately assess the degree of danger from just that one note? Can listening birds get sufficient information from a tiny snippet of sound to accurately decide whether or not they should flee to cover? And if they can, why give another string of notes if that first element can tell birds everything they need to know?" McLachlan suspects that the birds may be gleaning information from those additional notes about when it's safe to return to feeding. "There may be a cool dual mechanism for encoding urgency in alarm calls," she explains. "The acoustic structure of the first note may tell the birds whether or not to flee, and the number of elements in the call as a whole may tell them for how long they should stay in hiding."

McLachlan wasn't done. She suspected that the garden's smaller birds weren't the only ones listening in on the honeyeater's cry. "Compared with the usual flee alarm aerial alarm call, which is this high-pitched quiet thing that's hard to localize, the honeyeater's aerial alarm call is ridiculously loud, like they're calling out to *everyone*," she says. "I suspected this could be a signal to the predator as well—that it has been detected, its cover is blown, so there's no point in pursuing its prey.

"In biology, we tend to like dichotomies. Either it's a call to your conspecifics or it's a call to a predator. But I don't see any reason why you couldn't have a signal that's addressing both."

As a lark, McLachlan and her colleague Brani Igic went around presenting the honeyeater call to various birds. "Every bird we played the call to responded," says McLachlan. "Even the big 'lower end' predatory birds like kookaburras, ravens, and currawongs. They'd scan the sky or fly away."

So predator and prey alike are getting the gist of the honeyeater's alarm call message: "Raptor in the area, flee!" But do other species grasp the specifics, the detailed information encoded in one bird's call? In some cases, yes. Rob Magrath and his colleagues discovered that superb fairy-wrens and white-browed scrubwrens that live in the same area understand all the shades of meaning in one another's alarm calls. Australian magpies, which spend a lot of time on the ground, understand the encoded messages in the different alarm calls of noisy miners, which hang out in the trees. It makes sense for the magpie to attend to the warnings of birds with a better view.

Clearly birds are master eavesdroppers, listening in on alarms sounded by other birds.

So are other animals. In fact, one study showed that more than seventy species of vertebrates eavesdrop on alarm calls: Birds eavesdrop on other birds, mammals on other mammals, mammals on birds, and birds on mammals. In North America, chipmunks and red squirrels grasp the meaning of bird aerial alarm calls. In turn, chickadees understand the alarm squeaks of red squirrels and will take cover in response. Three species of lizards even attend to bird alarm calls. Yellow-casqued hornbills of West Africa can distinguish between the eagle and leopard alarm calls of Diana monkeys.

Does this mean there's a universal language of alarm? If not, how does one species understand the language of another?

"The long-held belief is that all alarm calls are acoustically similar enough to be innately understood by different species," says Magrath. That is, they all sound pretty much the same—scary, high-pitched, harsh, loud—and birds are born understanding them. Apostlebirds respond to the "foreign," but structurally similar, mobbing calls of an unfamiliar bird that lives halfway around the world, the Carolina wren. Swamp and song sparrows have distress calls that sound alike, and each understands the call of the other, but they don't respond to the distress call of a white-throated sparrow, which sounds different. The idea is this: Selection imposed by hawks may have resulted in convergence of an aerial alarm call—one that's high-pitched and subtle, and for a hawk, difficult to hear and to locate. This allows one species to understand the call of another. Mobbing calls, too, might have converged on that similar low-frequency, broadband, and repetitive sound to signal a terrestrial or perched predator. If that's true, it's easy to see how one species could listen in on another.

But Rob Magrath doesn't buy this as the only explanation. He says alarm calls vary a lot among species. The flee alarm calls of New Holland honeyeaters and superb fairy-wrens, for instance, differ in frequency, duration, and call structure, and yet fairy-wrens heed the alarm. "Historically, we thought bird alarm calls were like human screams, with a very noisy structure associated with huge emotion, attracting attention but not giving a specific message," says Magrath. "There was a general belief—there still is—that alarm calls are more or less easy to recognize because they're similar. But one of the early things we've found is that flee alarm calls and mobbing calls are in fact hugely diverse, and they're not necessarily automatically recognized by other species."

This led Magrath to the idea that cognition might be involved, that birds might have to *learn* the alarm calls of other species, just as they have to learn to fear—and mob—a novel predator. In the case of alarm calls, it might "almost be like learning another language," he says.

Do birds do this? Learn those calls the way we learn words in a foreign language? Or are they simply recognizing characteristics or qualities in the calls of other species? To find out, Magrath and his colleague Tom Bennett conducted experiments comparing the response of superb fairy-wrens to noisy miner alarm calls on the campus of Australian National University, where the miners are common, and across the road in the botanic gardens, where they're absent. When the team played the calls on campus, the fairy-wrens fled to cover. "And then we went across the road to the botanic gardens to see what would happen there," says Magrath. "We were fully expecting the miner alarm calls to have the same effect there because they sound pretty scary." But they didn't. The fairy-wrens didn't react at all.

Magrath is still astonished by the results. "If I step out of my door here on campus and do a playback to the fairy-wrens, they understand what's going on and flee. And if I walk five minutes up the road to the botanic gardens and do the playback, the fairy-wrens don't respond. To me, this is very startling. It means the fairy-wrens do not respond to unfamiliar alarm calls until they *learn* that they mean danger—in effect, until they learn the new language."

Learning, rather than familiar acoustic structure, determines a bird's response. "Birds learn who to pay attention to," says Magrath. "And this happens at an astonishingly fine spatial scale." (His paper is titled "A micro-geography of fear.") "That kind of flexibility is extremely valuable in a changing world, where individuals can be exposed to new species," he says—new predators *and* new informants. "Recognizing the alarm calls of lots of species is like understanding multiple foreign languages."

Could birds learn to link a new, unfamiliar sound with danger, like a made-up word or phrase? Magrath devised an experiment to teach wild superb fairy-wrens to associate a novel sound with the presence of a threat. The idea was simple: Train the fairy-wrens by broadcasting an unfamiliar sound while throwing over them a gliding model predator, a lifelike foam imitation of a currawong or sparrowhawk. Then, after training, play back just the sound to see if the birds fled. It sounds easy.

It wasn't.

The training task fell to Magrath's students at the time, Jessica McLachlan and Brani Igic. The setup was simple. Igic would hide in the shrubs with the model predator glider and wait. McLachlan, loaded up with playback equipment around her waist, would watch for birds, and when a fairy-wren appeared, signal to Igic to tell him when to throw the glider and simultaneously start the playback calls.

The problem was, the birds were wary. "We spent hours waiting for them to come out, to present," recalls Igic. And then the circumstances had to be perfect—fairy-wren out in the open, Igic positioned just right to throw the glider over the little bird. "And of course, when you throw a glider it doesn't always glide straight," says Igic. "Sometimes it hits a tree. And then, you're throwing these things and scaring the birds, and after a while, they become warier and warier, and they don't come out in the open much.

"We were in the gardens sunrise to sunset," says Igic. "Sometimes we would wait eight hours for one bird to come out, and then the window of opportunity would be so small that we would miss it. It was a simple experiment, very difficult to implement. Definitely the worst fieldwork I've ever done."

It's a good example of the extreme challenges scientists face studying bird behavior in the wild. "You can understand why people who study learning do it in boxes, on rats and pigeons," says Magrath—"because you can control everything. You can control the bird's experience. It has to stay there. It can't hide in the bushes. Doing things in the field is much more difficult—even if the questions are very simple."

The grueling work paid off. Igic and McLachlan managed to repeat the training process of throwing the glider while playing the sound eight times to ten fairy-wrens, and the fairy-wrens mastered the new vocabulary in just two days. Says Magrath, "I remember Jess and Brani bursting through the door, saying, *It worked!*" Before the training, the fairy-wrens ignored the novel sound, but after training, they fled to cover just from playback of the new alarm sound, without the presence of the model sparrowhawk.

Wild birds had succeeded in learning a new language.

It makes sense that birds are quick studies, good at gleaning lessons from a single experience, especially where danger is concerned. They can't afford to forget. As behavioral ecologists have drily noted, "Few failures . . . are as unforgiving as the failure to avoid a predator: being killed greatly decreases future fitness."

Eberhard Curio's blackbirds showed us that birds don't necessarily know innately which predators to fear. They learn from either direct experience or social learning. To study how birds learn about predators, biologist Blake Carlton Jones set up an experiment to see if a bird could learn about a novel threat after only one encounter. His "predator" was a black umbrella with big yellow eyes. At first, the birds showed no interest or alarm. But after the umbrella chased the birds just once for five seconds, they remembered the encounter, even four years later, and when the umbrella appeared, they instantly fled. No repeat lesson needed.

More recently, Magrath has found that fairy-wrens can learn unfamiliar alarm calls of other species just by listening and making associations between two sounds. This is a good example of social learning—like Curio's mobbing experiment—learning from other birds rather than through direct experience. Magrath trained the birds by broadcasting novel sounds together with a chorus of the familiar alarm calls of fairy-wrens and other species. Before training, the birds didn't flee from the new sounds, but after training, they did. They didn't even need to see the predator themselves or other birds fleeing it. The sound association was enough.

"Which means, theoretically, they could learn with their eyes closed!" says Magrath. "I like the idea that they could do it solely through acoustics, because that's something birdwatchers do, too, 'watching' just by listening."

Learning by sound association helps birds with the extraordinarily complex task of sifting the acoustic world for cues about danger. "Imagine in a rain forest, you can hear an alarm call but you most likely can't see the predator, who's fast and elusive and may be upon you before you know it," says Magrath. "Through learning, a bird can potentially use a diversity of alarm calls given by the local bird community, taking advantage of a vital information web. Learning socially, learning from others, spares birds the risk of direct personal experience." It also allows the knowledge—and the presence of informants—to spread.

McLachlan has a microphone going the whole time she's in the field so that she can opportunistically record alarm calls whenever they occur. "They're so unpredictable," she says. "I don't know when they're going to fly through, so I just need this thing recording all the time." The downside of this, she says, is that her vocalizations are also recorded. She decided to take advantage of this and do a little experiment on her conspecifics, the members of her family. She selected a batch of recordings that captured her vocal reactions to things in the garden that startled her, to see what information her family member(s) could glean from *her* vocal alarm calls.

"During one playback experiment, I noticed there was something on my shirt," McLachlan recalls. "But I had my eyes on the camera and told myself, 'I'm filming an alarm call, it really doesn't matter what it is.' And then it moved a bit. But I was sure it was a leaf or something, and I told myself, 'It's fine, just don't care, don't care.' And then as I finished the film, I looked down, and saw it was a huntsman spider, and it was climbing up my chest. According to the recording, I just screamed, '*Ahhhhhhh!*' which I would like to claim is inaccurate, but it's on tape."

The huntsman is a hairy, scary spider, *huge*, with a leg span of five inches—an arachnophobe's nightmare. But at least its venom is not very dangerous.

Which is not something you can say about the eastern brown snake McLachlan nearly stepped on. According to *Australian Geographic*, this snake is "fast-moving, aggressive, and known for its bad temper, responsible for more deaths every year in Australia than any other snake." Its venom, ranked as the second most toxic of any land snake in the world, causes progressive paralysis and stops blood from clotting. Victims may collapse and die within minutes.

"I'd almost put my foot right down on one while filming," says McLachlan, "and I thought *that* recording would be an entertaining one to send to my family. I knew what I'd said, pardon my language, 'Holy shit!' I thought I'd played it fairly cool. But then I listened to the recording again. It was more like 'Ho-ho-ho-ho-*holy shit!*'"

McLachlan sent the recordings of her vocal expressions of alarm to her family members and asked them what information they could extract from these alarm calls. "Can you tell me the category of threat? What kind of animal? Is it in the air or on the ground? How close by? How fast is it moving?"

All they got was her fear.

"We have this really sophisticated language to communicate extremely detailed concepts," says McLachlan, "but when we're communicating urgently about danger, we don't use helpful words that describe what it is, where it is, how close it is—nothing useful."

That birds can pack so much information into their calls—type of predator, perched or in flight, near or far, how fast approaching, how dangerous—perhaps should not be so surprising.

For millennia, humans have used languages consisting solely of whistles to organize, argue, gossip, even flirt across distances. We think of whistles as a way of getting attention or carrying a tune, incapable of conveying much meaning. But go to Antia in Greece, the foothills of the Himalayas, the Canary Islands, the Bering Strait, Ethiopia's Omo Valley, the Brazilian Amazon, and dozens of other remote places, and you may hear volleys of human whistled chirps, cryptic trills, and fluted duets, entire conversations of lilting whistles a lot like bird sounds that communicate with all the subtleties of human speech.

Julien Meyer of the University of Grenoble, France, has studied whistling languages for decades and has identified as many as seventy groups that use the language, most often in mountainous areas with steep terrain or in dense forests. Whistlers make their sounds in different ways, he says, sometimes by cupping their hands in front of their mouths, forming a resonant cavity, or rotating the tip of the tongue against the lower teeth, or putting two fingers in the mouth, or holding a leaf between the lips. In all cases, the result is a melody that imitates the syllables of ordinary speech. Whistled sound travels about ten times farther than spoken words, up to five miles in open places, and can penetrate the thick foliage of rainforest, so it's useful to the shepherd in the mountains, the hunter in the forest, the farmer in the steep ravine. In the Canary Islands, the whistled form of spoken Spanish is known as Silbo, and the trills shepherds use to converse across ravines closely resemble the song

of local blackbirds. In the small Turkish town of Kusköy, the local whistling language is called *kuş dili*, or "bird language."

All of these whistled languages are based on the elements of spoken
language. In some, "vowels" may be represented by different types of
resonance in the vocal tract, resulting in different notes, and "consonants,"
by sudden leaps or slides from one note to the next. A series of five syllables in Turkish, Greek, or Spanish becomes five different whistles made
with the teeth, tongue, and fingers. A fluent whistler can decode a "sentence" of tweets, warbles, and trills with 90 percent accuracy for simple
sentences, about the same as speech. The languages are used to organize
daily activities, announce an emergency, or transmit news or secret or
private information, says Meyer. In parts of Southeast Asia, it's used to
convey love poems.

No one knows how these whistled languages began. In Greece, one
theory holds that whistlers were posted as sentinels on mountaintops so
they could warn of impending invasion or attack.

Sound familiar?

If humans can encode all those specific messages in whistles, why not
birds? The complexity we've so far decoded in bird alarm calls makes
me wonder what else we are missing. Scientists who study bird calls wonder this, too. The main research tools for analyzing birdcalls are, in the
words of avian vocal experts at the Cornell Lab of Ornithology, "relatively crude ways to picture sounds, showing only selected features of
this extremely rich form of communication."

What are birds really hearing? And what are they attending to? We
don't process sound as rapidly as birds can. They're far better than we are
at distinguishing minute differences and quick changes in the structure
of acoustically complex songs—tiny shifts and differences in the qualities of a sound inaudible to us. They're capable of picking out separate
sound sources much faster than we can and of processing many more
notes in a given time period. The ways we interpret and categorize bird

vocalizations may bear little resemblance to how birds actually hear and use them.

McLachlan says she wonders about how birds classify threats. Is it really all about aerial versus ground or perched predators? Or might predator speed or angle of approach factor in? "When a kookaburra flies down to grab something on the ground, prey birds give aerial alarm calls," she points out, "but then when it flies much more slowly back up, they switch to mobbing calls. The kookaburra is still in flight, but both its speed and angle have changed."

Rob Magrath has found that he can prompt fairy-wrens to give aerial alarm calls if he bicycles by them really fast. "So again, it could be speed that they're paying attention to," says McLachlan. She has also prompted their aerial alarm calls at night by quickly flashing a torchlight around.

Most exciting to McLachlan is some intriguing new work suggesting that songbirds possess a language-like ability to combine vocalizations, creating more complex meaning. "This is where we're really getting a view of what's going on inside a bird's head," she says.

We have long thought of language as a bellwether of human uniqueness, the most important feature that separates us from other animals. And it's true—no other creature apart from humans can take a limited selection of components—more than ten thousand words for most adults—and arrange them in a vast array of combinations, each with a precise meaning. But science is finding more and more parallels between speech and bird vocalizations, as well as language-like qualities in bird songs and calls.

Alarms calls are a prime example. Not only are they functionally referential—working like words do to describe the characteristics of an object—but according to a few recent studies, they may also show some key linguistic features of human language. Compositional syntax, for instance, the set of rules for arranging and combining sound and words to create meaningful phrases and sentences. In language, it matters how we order our words—and how we combine them to form more complex messages. In English, *watch out* makes sense; *out watch* does not.

We thought only humans used syntax in this way. But Toshitaka Suzuki has found that Japanese tits seem to use these sorts of linguistic rules in their alarm calls. The tits have a repertoire of eleven different notes, which they use alone or in combination with other notes, resulting in more than 175 different call types, and listening tits respond accordingly. An "ABC" call appears to warn other Japanese tits to scan for predators. A "D" call (like a chickadee's *dee*) urges other birds to come hither. The birds join these two calls into an "ABCD" call when they're recruiting other tits to collectively mob a stationary predator, such as a marten. "Tits combine different meaningful signals to generate a compound message that depends on the meaning of the elements and the way they're combined," says Suzuki. Switch the order of the calls, and the meaning of the call changes.

Other birds may have language-like abilities, too. Southern pied babblers have the same ability as Japanese tits to combine meaningful signals into a meaningful sequence. These highly social birds of southern Africa spend their lives probing the desert sands for hidden invertebrates. Most of the time, they're in a head-down position and have to rely heavily on an array of vocalizations—alarm calls, sentinel calls, and recruitment and other social calls—to keep track of what's going on around them. Sabrina Engesser, an evolutionary biologist studying bird communication, has found that when babblers encounter a land predator, they combine their distinctive "low-urgency threat" call with their recruitment call to form a mobbing sequence, which recruits group members to the potentially dangerous situation. If a large raptor shows up that needs mobbing, such as an eagle, they'll combine an aerial alarm call with a recruitment call. Young babblers combine begging calls with recruitment calls when they're out foraging with foraging helpers. Change the order of the calls, and the babblers will not respond. The birds are linking separate calls (alarm calls or begging calls) with command coordination, just as we do, say, in the warning, "Danger! Come here!" "They perceive these complex signals as compositional structures made up of discrete signals," says Engesser, "which, when combined, encode a message that goes beyond the meanings of its individual parts."

Engesser and her colleagues have also discovered that chestnut-crowned babblers, a cooperatively breeding species found in southeastern Australia, combine meaningless sounds to create meaningful calls. In linguistic terms, this is akin to using phonemes to create meaningful words—for example, *p* and *u* in isolation don't mean anything but can be combined to form *up* and *pup*. "What these birds do is in a very rudimentary sense akin to word creation," says Engesser. "What pied babblers and Japanese tits do is more like the creation of a phrase or short sentence."

According to Suzuki, Japanese tits can even use grammatical rules to decode novel combinations of calls. Suzuki created a call that these birds had never heard before—an artificial sequence that combined their own ABC alert call with the *tää* recruitment call of another species, a willow tit. When he played the combination calls, the Japanese tits decoded the meaning of the ABC-*tää* combination, but not when it was played in reverse. This suggests that the birds are not just memorizing complex signals but rather applying a generalized grammatical ordering rule to decode messages.

Not everyone swallows the notion that bird vocalizations show the kind of compositionality you find in human language. As many linguists argue, all known cases of syntax in birds involve combinations of only two meaningful calls, while human language can combine different words into an infinite number of expressions. Still, as Suzuki says, his latest study suggests an interesting parallel with grammatical ability in human language and may shed light on its evolution.

There may be other ways in which a bird's use of call and song mirrors our own use of language and reveals intelligence—for instance, in its powerful capacity to deceive and manipulate. As one psychologist said, "The truth comes naturally, but lying takes effort and a sharp, flexible mind."

Chapter Three

SUPERB PARROTING

The forest smells cool, damp, and green. Beneath the shade of giant eucalypts, the trail is lush with low-growing bracken ferns and ancient six-foot-high tree ferns, the ground muddy with flowing streams. This path, the Myrtle Gully Circuit trail, lies deep in the Toolangi rainforest in southeastern Australia, a primeval place with towering myrtle beech and mountain ash trees. Among the world's tallest hardwoods, the ash is also known for its distinctive bark, which grows fibrous at the base but thirty feet up yields to a smooth mottled surface of pale cream and gray, then to streamers of dead bark ribboning from the high canopy. In the understory beneath the canopy, the air is still and moist, and the light, almost emerald. A loud staccato twittering pipes through the eucalypts. My guide and friend, Andrew Skeoch, instantly identifies it as a honeyeater, an eastern spinebill. Then the metallic *Egypt!* and *choc-chip!* calls of the crescent honeyeater, the soft *churr*s of a brown thornbill.

Andrew knows his birdcalls. An expert in wildlife sound recording, he has been capturing the sounds of nature since 1993. Toolangi was the site of one of his earliest professional recordings. Since then, he has made recordings all over the world, in India, Turkey, Thailand, Sweden, Malaysia. Some recordists aim to catch the signature songs and calls of a single species. Andrew strives to capture the acoustic signature of a whole habitat. "There's so much to hear," he says. "Just by listening, you can recognize what species are around, what they're doing, their interactions. You can hear the actual *process* of an ecosystem." Here, the sound is liquid, tinkling, and echoey with the collective trills, peeps, chips, and whistles of birds.

On this expedition, however, we are seeking a singular species, the superb lyrebird, Australia's centerfold singing bird. This fern gully is ideal habitat, but the bird is shy and elusive. We listen hard. In the dim cloak of this rainforest, you hear things before you see them—the ringing *kling* of a grey currawong and the *chack chack* of a king parrot. A crimson rosella jets through the midstory, uttering its harsh *klee-klee-klee*. It alights in a massive eucalypt and switches to a ringing bell-like *po-ti-po* call—"close to what we recognize as the musical interval of a fifth," Andrew says, from *do* to *sol* to *do*. The loud piping call of the diminutive white-throated treecreeper reminds me of the big bold sound of our Carolina wrens.

Then something else pierces the quiet green. A far-off *pssew, pssew,* followed by a stream of weird, wonderful, haunting sound.

Andrew raises a finger. "That's it." The distant song of the lyrebird, surely one of the most extraordinary voices in the animal kingdom.

The superb lyrebird seems a modest bird, coppery brown and pheasant-like, until it raises its tail or opens its throat. Spectacular tail feathers shaped like a lyre—two long, curved, outer feathers forming the arms, a set of white filamentous feathers, the strings—give the bird its common name. But its voice is what makes it truly superb. The lyrebird sings like no other bird, a fantastic blend of its own calls and songs and dozens of perfectly mimicked sounds, brilliant imitations of other bird voices in the forest, from the explosive "whipcrack" of the eastern whipbird to the ringing *pee-o* of the grey shrike thrush, from the stuttered creaks of gang-gang cockatoos to the silvery sweet chant of the pilotbird.

"Oddly enough, the lyrebird doesn't mimic some of its most common neighbors, such as spinebills," says Andrew. But it will faithfully parrot even the raspy scold and buzz of the tiny white-browed scrubwren and the high urgent chittering of the little brown thornbill—birds with bodies one-seventieth its size.

The muddy trail narrows as it climbs gently toward the ridge until ferns are brushing us on both sides. Fortunately, it's past leech season. On the moist forest floor, every fallen branch and log is blanketed with mosses, liverworts, fungi. Even the live tree trunks soften with their pelt of epiphytic mosses known as old man's beard.

Everywhere along the path are tantalizing signs of recent lyrebird activity, scratchings in the soil and leaf mold where the big bird has foraged for insects, worms, and grubs with its powerful dark feet. It stands on one leg, scrapes with the other, head perfectly still. In this way, lyrebirds can shift tons of soil and litter in just a year.

Singing males have been recorded along these slopes at regular intervals, every four hundred feet or so, Andrew tells me. But that's in the peak of the breeding season, June and July. It's mid-August now, late in Australian winter. "If we were here a month or two earlier, we'd be having our eardrums rearranged," he says. The lyrebird's song can be extremely loud, painfully so if heard at less than thirty feet.

Andrew stops again. "I think I just heard one doing a yellow-tailed black cockatoo. That eerie squealing, *kee-ow, kee-ow.* Unless that's actually a cockatoo."

He pauses.

"Yes, it's a cockatoo!" He gestures wildly at the nearby midstory as the large black bird John Gould called the funereal cockatoo wheels lazily past, flashing a tail patched with brilliant yellow.

We hear another lyrebird singing from far across the creek, a distance of a half mile or more. It seems the lyrebirds are teasing us, staying just a few steps ahead or watching us pass from a cloak of ferns, always just out of sight.

By the time we reach the top of the ridge, it's late in the day. We have the option to take the easier, drier descending leg of the looping trail, but Andrew thinks we have a better chance of seeing this furtive bird if we go back down the way we came, through the mud. I reluctantly agree, and we slog downward along the trail. Not far from here, Andrew and his partner, Sarah, once found a female lyrebird hanging inverted from a branch, with one toe caught in a vine. Andrew took off his shirt and wrapped it around the bird to support it, while Sarah untangled its toe. "The bird's eyes were locked on me the whole time," he recalls. "As soon as I put her down, and her feet touched the ground, she was off! Maybe the rescue will bring us good luck today."

Maybe. But I'm losing hope. The mud is soaking through my boots and jeans, and the afternoon light is waning. Evening descends quickly in the rainforest, drops like a curtain. We still have a long way to go.

We pause for a drink of water.

Then suddenly we hear it. A powerful burst of song not thirty yards from where we stand. We freeze and listen. It's a glorious performance, booming, resonant, with a splendid cascade of melodies, mingling the bird's natural notes with stolen calls and songs. Andrew whispers out the species. *Whipbird. Rosella. Grey currawong. Pilotbird. Pied currawong.* I'm glad he's by my side translating. Never have I thrilled to birdsong so fluid, pure, and orchestral, so packed with borrowings, as if all the songs of the rainforest were pouring from a single throat. Now I understand why this bird is called the mother of all songbirds.

We creep closer and catch the quickest glimpse of the lyrebird's light, shimmery white tail. Then it's gone, along with the marvel of its operatic song.

As we descend quickly through the darkening forest, bobbing about in my skull is a host of questions. Does the lyrebird mimic just some songs and not others? What kind of intelligence does it take to mimic— and to mimic well? And why in the world would one bird bother to imitate the voices of so many others?

To explore these questions, I visit Ana Dalziell, an expert on lyrebirds and avian mimicry, now at the University of Wollongong in New South Wales. More than any other scientist, Dalziell has probed the

complex and sophisticated vocalizations of the lyrebird, including a puzzling propensity to mimic alarm calls while it's mating. Her research
goes to the heart of a deep mystery in bird talk: the nature and purpose
of mimicry and how birds may use it to profit themselves and, perhaps,
to manipulate and deceive others.

Until recently, we thought humans had a lock on deception and lying.
We know we're good at it. Researchers recently investigated the ubiquity
of lying in day-to-day life in the United States and found that the mean
number of lies told on an average day is one or two. This seems low to me.
And maybe it is: These were self-reports, which raises the very real question of whether people tell the truth about lying. But most people do seem
to opt for honesty and turn to deception only when they feel they have to.
This is probably because for most of us, lying is hard mental work.

This is one reason we thought other animals were incapable of it.
Lying is a sophisticated behavior, says biologist Robert Sapolsky. When
you lie, you understand that your message is false, and that the recipient of the lie will believe your falsehood. That means understanding that
your belief differs from theirs, that they have a mind with knowledge
that differs from your own—a capacity known as theory of mind. Lying
also demands significant effort by the brain's executive processes: attention, planning, inhibition, working memory. It requires conscious and
sophisticated strategizing, and its complexity is reflected in the little
ways our facial muscles, innervated as they are with huge numbers of
neurons, so often betray us. Like most mentally taxing activities, however, it gets easier with practice.

When scientists studying deception in primates compared different
species of monkeys and apes, they found a direct correlation between
guile and brain size. Primates with more neocortex, such as chimpanzees, are more likely to pull off deeds of deception. Take the famous
chimp Santino in a Swedish zoo who stockpiled rocks to throw at visitors
when he was in a foul mood. When zoo workers cottoned on to the stunt,
they removed hundreds of caches and told visitors to back away when
Santino held a projectile. The chimp, in turn, learned to hide his caches

of rocks in piles of hay and sticks and to conceal his aggressive displays, which were telltale signs of upcoming throws.

Birds are among nature's most gifted liars, capable of just this sort of nuanced deception, says Dalziell. The bird world is rife with bluffs, masquerades, shams, and shell games. Some parent birds, such as piping plovers, feign a broken wing to draw predators away from the nest, fluttering erratically and making convulsive attempts to run, jump, or fly. Other birds distract predators by running in a crouched position like a small rodent. Still others, such as quail, feign death to fool their pursuers. Birds that cache their food, such as scrub jays, will move their food stashes several times if they know they're being watched by another jay, shifting it to different locations or even fake-moving it but leaving it buried, a shell game aimed at confusing the viewer. These are all instances of physical deception.

But do birds use their voices, their mimicry, to deceive? And to what end? Can a lyrebird lie?

On a cool August morning, I travel northwest from Sydney by train to join Dalziell on a trip to her research sites in the Blue Mountains of eastern Australia. It's a clear, late-winter day with a strong wind blowing—almost a gale. From the train I can see flocks of sulphur-crested cockatoos wheeling above the countryside and galahs floating by like little perpetual sunsets. Just past Emu Plains, a wall of green rises, the eastern escarpment of the Blue Mountains. Dalziell picks me up in her Land Cruiser and we drive to the sites in a hidden valley in the mountains. She warns me that the male lyrebirds tend to be quiet and reserved from August through September, having just finished their peak May to July singing, and are busy growing a new tail for the next breeding season. But she's happy to discuss their nature and to show me their "quite spectacular" habitat in the fern gullies and wooded sandstone gorges of the Blue Mountains.

"Quite spectacular" is putting it mildly. These old mountains are crumbling away, leaving dramatic rock formations, triads of sandstone pillars that spring out of the dense, almost fluorescent green of the forest, and sheer cliffs that shelter little pockets of temperate rainforest in the valley below. Dalziell's study site is so remote that a rare species of tree, the Wollemi pine, was only recently discovered not far from there. It's fine lyrebird habitat, cool and wet, with creeks threading the forest, and abundant food.

Dalziell is disturbed by the wind howling around us. "We're unlucky," she frets. I assume that she's concerned that the high winds will keep the lyrebirds in hiding. "That, too," she says. "But mainly I'm worried because there's a long history of people dying from trees falling in these forests. These are some of the continent's tallest trees, and they lose their limbs the way North American trees lose their leaves. Around here, the eucalypts are called widow-makers."

At Kedumba Pass, we drive through a locked gate into more remote, restricted territory. Dalziell travels with a radio and an EPIRB, an emergency position indicating radio beacon. She texts the park ranger to let him know exactly where we are.

Sheer cliffs tower to our left. We wind along the base of the cliffs, wind thrashing the eucalypts. Another sharp curve in the road, and it suddenly quiets; then the road curves again, and the wind is wilder than ever. It has been a hard year for lyrebirds, Dalziell tells me, weirdly warm, with frequent high-intensity fires. The ecosystem here is well adapted to fire, but huge once-in-a-generation fires are now happening more often. Just after Dalziell moved to the town of Winmalee early in spring, a big fire fueled by bone-dry, howling hot winds swept through the area and nearly engulfed her house. She worries about lyrebirds in the face of climate change, especially the extreme heat and fire events in winter, when they breed. The birds survive the fires by going down a wombat hole or slipping into water. There are stories of people seeking refuge from a fire in a creek, pulling a wet blanket over them, and finding themselves next to one very fancy bird.

A rockwarbler hops across the road, springing from its feet as if somehow defying gravity. In the sheltered areas of the cliffs is wet sclerophyll forest, a moist woodland of eucalypts, wattles, and banksias, deep, dark, and green with blanket leafs, stringybark, and she-oaks, as well as rock orchids so valuable that people have been known to steal them. "But they're safe in this area," says Dalziell. "Hardly anyone gets here." A brown thornbill calls, then a yellow-throated scrubwren and a golden whistler, a beautiful bird with a beautiful voice—a brisk *sweetaWIT*, *sweetaWIT*. All species the lyrebird imitates.

Dalziell is fascinated by avian mimicry. "The mimetic feats of birds can be gobsmacking," she says. "I mean, the little brown thornbill, which weighs just six or seven grams, can imitate the song of birds more than ten times its size." The red-capped robin-chat of Uganda mimics the sounds of forty other bird species, including their high-speed duets. The marsh warbler has no original song of its own but weaves together song elements from up to two hundred different species that live on its wintering grounds in Africa and its breeding grounds across Europe, combining and reconfiguring them into a fantasia of stolen song torn from their musical context.

It's not known how many bird species mimic. One database of avian mimics lists 339 species in 43 different families of songbirds. It's rare in other bird groups. Parrots are highly prized for their ability to imitate the human voice in captivity but reports of parrot mimicry in the wild are rare, except possibly in African grey parrots. Sometimes mimicry pops up only sporadically in a bird's repertoire; sometimes it accounts for nearly all of its vocalizations, as in male marsh and icterine warblers. Australia is known to have some of the best avian mimics—perhaps as many as 60 species, among them bowerbirds, mistletoebirds, silvereyes, olive-backed orioles, scrubwrens, and thornbills. Fourteen species have been described by early naturalists as "master mimics," including Australian magpies, birds that "adopt and adapt whatever their fancy dictates," as one ornithologist wrote, "blend the borrowing with their own and make the whole performance so artistic that it becomes completely personal." Andrew Skeoch recorded one Australian magpie neighing exactly like a horse.

Among the accomplished vocal imitators in North America are the catbird, the northern mockingbird, and the brown thrasher, which is said to sing as many as two thousand different songs. European starlings imitate several other birds, along with everything from car alarms and cell phone rings to barking dogs. The blue jay's ability to imitate a hawk is so good that I've often looked skyward vainly searching for a red-tailed hawk wheeling above, when the source of the hawk's keening call is a blue jay in the nearby understory.

However, says Dalziell, mimicry in lyrebirds is unsurpassed. The birds have their own calls, a repetitive *plik-plik* that sounds like dogs barking or the blows of an ax, along with clicks and clucks and rhythmic thuds, like the hooves of a trotting horse, croaks and cackles, scoldings and squawkings and sweet melodic warblings, mechanical whirs, and strange, spectacular twanging sounds. They also utter their own intense alarm call, a piercing, high-pitched, screeching *whisk!* But mimicry dominates their repertoire. Both males and females mimic, explains Dalziell, but the sexes differ in the species they favor, with females imitating more raptors, such as the collared sparrowhawk and the grey goshawk and the mobbing calls of the territorial bell miner. Males switch between sounds quickly, says Dalziell, specializing in sharp contrast, "two seconds of eastern whipbird, then two seconds of crimson rosella, followed by three seconds of grey shrike thrush." Around here, the males have what she calls a set of "greatest hits": kookaburras, whipbirds, yellow-tailed black cockatoos, golden whistlers, and satin bowerbirds, which are great mimics themselves.

"The lyrebirds are mimicking mimickers," says Dalziell. "And they're wonderful mimics of their own kind. It can all get very complicated for researchers trying to do a bird transect, surveying and recording the variety of species in an area. They think they're hearing a particular species but it's actually a mimic imitating that species." I think of Andrew's confusion over the call of that yellow-tailed black cockatoo.

Are their impersonations good enough to fool the birds they imitate?

According to Dalziell's research, yes. The grey shrike thrush, for instance, is often duped by the lyrebird's impression of its own song, even though the lyrebird delivers a rushed and abridged—what Rob Magrath calls a *"Reader's Digest"*—edition of the beautiful, leisurely shrike thrush song.

Lyrebirds can even mimic several calls at one time: the duets of whipbirds, the giggling, tinkling chatter of a flock of parrots. Dalziell once recorded a lyrebird mimicking a whole chorus of laughing kookaburras.

Despite the popular impression that the birds frequently imitate human-generated noises—perpetuated by David Attenborough's impressive show of a lyrebird mimicking with amazing fidelity a chainsaw, fire alarm, and camera motor drive—Dalziell says that imitation by wild birds of human-made sounds or introduced species is extremely rare. There are, however, reports of lyrebirds mimicking random natural sounds: frog and koala calls, the howling of a dingo, a cockatoo's shredding of wood. Dalziell has recorded them imitating the wingbeats of birds fluttering through the forest understory. And one female lyrebird in her study area is fond of making a sound we heard often that day, the distinctive *squeak, squeak* of tree trunks rubbing together in the high winds.

The lyrebird's broad vocal range is surprising given the reduced number of muscles in the lyrebird syrinx, its vocal organ. Instead of the usual four muscle pairs possessed by most other songbirds, lyrebirds have only three. Typically, songbirds with more muscles in their syrinx can generate greater vocal complexity. But the lyrebird is an exception to this rule. So are parrots, which also use a simplified syrinx to produce all kinds of skillful sounds.

Mimicry was once chalked off as mindless behavior, a belief reflected in our verb *to parrot*, with its overtones of mechanical repetition without appreciation of meaning. Now we know it's anything but.

Imitating a song or call requires vocal learning—listening closely, memorizing, recalling, and practicing—trying repeatedly to produce a perfect imitation of what is remembered and making corrections until the copy is a match. Learning and remembering a sound made by another species is especially demanding—requiring a brain flexible enough to acquire a new sound not part of the normal repertoire of a species—and so is the physical task of accurately reproducing that sound. Some bird mimics somehow engage the same complex control mechanisms of the muscles in their syrinx that are used by their models. In 2008, researchers at Duke University studying swamp sparrows discovered that the birds use "mirroring neurons" when they're learning vocalizations, just as primates do. When one bird listens to another bird's song, its neurons fire in nearly identical patterns of activity as when the bird sings the same vocalization itself. Imitation, then, is a way of connecting one mind to another.

All of this requires sophisticated and superb neural functions—good hearing, memory, and consummate muscle control for sound production. That lyrebirds excel at the task suggests exceptional intelligence. Indeed, when scientists compared the brain volume of various birds relative to their body size, they found that lyrebirds have very large relative brain size, just as corvids do. The birds also develop slowly, with a long incubation period and an extended period of parental care, which is also linked with higher cognitive function.

How birds learn their mimicry is still largely a mystery. Some mimics appear to learn directly from the species around them. Not long ago, Laura Kelley, then at the University of Edinburgh, looked at learning in the spotted bowerbird, a highly gifted mimic, which typically imitates more than a dozen other species, as well as the creaking of branches and twanging of wires, even the rolling of thunder. Male spotted bowerbirds build bowers at least a kilometer apart, so they're unlikely to hear one another's vocalizations when they're in their own bowers. However, they do regularly visit neighboring bowers to steal their decorations or destroy them. Do the birds learn from one another during these sporadic visits? When Kelley investigated the bowerbird's mimicry of whistling kites and pied butcherbirds in a national park in Queensland, she found that the bowerbirds showed individual differences in renditions of their mimicry, suggesting they were learning not from their neighbors but directly from their models.

On the other hand, northern mockingbirds seem to learn their mimicry from other mockingbirds. Not long ago, biologist Dave Gammon introduced the songs of eight new species to mockingbirds over a six-month period, and none of the mockers picked up the new songs.

The question of why mockingbirds mimic certain birds (Carolina wrens, tufted titmice, blue jays, cardinals) and not others (mourning doves and chipping sparrows) may shed light on the lyrebird's selective mimicry. The mockingbird mimics only those songs similar in pitch and rhythm to its own vocalizations. It may be physically unable to replicate certain features of other species' songs, such as smoothly coordinated switches from one side of the syrinx to the other or jumps in the frequencies of notes. This may be true of lyrebirds, too.

Dalziell thinks that young male lyrebirds, like mockingbirds, learn most of their mimicked calls from older adult males rather than directly from the species they imitate. "Their mimicry seems to be in the fashion of other males in the area," she says. "There's less variation within a population and more variation between populations." Male lyrebirds living in the same region mimic the same combination of calls and songs and even the same idiosyncratic combinations of sounds, she says, which is very suggestive of social and cultural transmission. In other words, their mimicry is not contemporaneous with what's around but is passed along from bird to bird. "For example, in this pocket of rainforest, a number of males mimic the first few notes in the songs of the white-throated treecreeper, a rapid piping *beep-beep-beep*," she says, "and then end their song with the raucous *bonn-bonn* of the crimson rosella." Evidence for this social diffusion of mimicry comes from a story told often about a group of twenty lyrebirds caught between Toolangi and Warburton and introduced to the island of Tasmania between 1934 and 1949. The birds continued mimicking birdcalls from their old landscape for many years. Thirty years after they were released, their descendants were said to be imitating birds never present on the island, such as pilotbirds and whipbirds. To Dalziell, this represents compelling proof of cultural transmission, one generation passing on knowledge to the next. In fact, she describes lyrebirds as archivists of soundscapes—the bird version of Andrew Skeoch.

In an area sheltered from the wind, Dalziell parks the Land Cruiser at what she calls a lyrebird "hot spot." We start upslope through a thicket of ferns to look for the nest where a year ago she planted a camera to monitor lyrebird breeding activity. Dalziell warns me not to touch any spiders. The bite of the Sydney funnel-web spider found around here can lead to profuse salivating, sweating, muscle spasms, hypertension, elevated heart rate, heart attack, and death in fifteen minutes.

At least it's too cold for snakes. We strap on gaiters anyway. The ground is littered with fallen logs and boulders hidden by the thick ferns, and the footing is treacherous. Prickly smilax tears at our jeans and grabs our sweaters. We find a scat pile from a swamp wallaby. "Amazing to think that wallabies navigate this landscape with their big feet," Dalziell says. Everything around us is spiky to protect against their grazing, and Dalziell wonders aloud how the male lyrebirds get through the smilax without shredding their long, elegant tails. They molt them every year right after the breeding season, immediately grow new ones, and somehow keep them intact until the following June.

Up the slope we find a lyrebird nest perched on a boulder beneath the crown of a tree fern and—to Dalziell's relief—her camera, still in one piece, nearby. The bulky nest is round, like a basket inside, and made from the tightly woven rootlets of tree ferns.

Look up, says Dalziell. It's almost completely dark above, the better for the female to protect her young from a host of aerial predators— goshawks, sparrowhawks, wedge-tailed eagles, as well as butcherbirds, magpies, and currawongs. The nest faces downhill so that the birds can escape by leaping or flying downslope. They are poor fliers but can run up to twenty-five miles an hour and use their wings to increase their speed. A female lays just one egg per season, incubates it for as long as two months—a world record for songbirds—and cares for her young all by herself, feeding it from her territory, which she defends fiercely against other females.

Not far from the nest, in the middle of a patch of tree ferns, we find a cleared area about three feet in diameter, its soft soil surface slightly elevated, oddly luminous. A stellar example of a male lyrebird's display mound, Dalziell says. The mound is the result of hard labor. The birds use their strong legs to rake the area bare, kicking aside small boulders and branches, yanking up ferns and dragging them out. "A male will build more than ten of these in his territory," said Dalziell, "but he has his favorites. Usually on high ground, where the canopy is not as dense, light can penetrate, and calls can travel far.

"Look up again," says Dalziell. We see sunshine.

Dalziell hypothesizes that the male birds perform in patches of sunlight where there's a break in the canopy to maximize the brightness of their white tail feathers. And because lyrebirds may be more sensitive to light than we are, she suspects that the bright dazzle is more intense for them than it is for us. Putting his site in a good spot is a big part of the success of a male lyrebird's display, and he will fight off other males to defend his mounds. Here's where his mimicry will shine in the breeding season.

In the still, misty days of winter, a male starts off singing in the early morning, a half hour before dawn, and will go on singing for hours each day, day after day, showing off his skills to females he can't see in the dense forest around him—until he finally draws one in. If she shows interest, she's treated to one of the world's most bizarre mating displays. Dalziell and her colleagues caught on video twelve of these performances by males in their natural habitat—not an easy task given the lyrebird's secretive nature and the green gloom of his forest home.

The videos reveal the details: The male bird on his mound sweeps his silvery tail up over his body, turning slowly around, shivering it theatrically and pouring forth a stream of song and mimicry. He flaps his wings and hops from side to side, all the while emitting songs, calls, castanet-like clicks, chirps, and chatters from his vast repertoire. He dances to only four types of songs, says Dalziell, and each song type is accompanied by its own moves. "He'll coordinate his song and dance so that each different song has a unique choreography." Just as humans waltz to waltz music and salsa to salsa music, she says, "so a lyrebird will dance a side step in time with a weird buzzing *spew, spew, spew* that sounds like a laser gun or a 1980s video game, and then, with his tail narrowed and wings flapping, jump or bob deeply while singing his more quiet *plinkety-plinkety-plinkety* song."

Then, at the very end, he sometimes does something mysterious—abruptly breaks his song and dance routine and launches into a series of mimicked alarm calls.

Why birds mimic at all is still something of a mystery. Why go to all the trouble of impersonating other birds? Listening to the lyrebirds' melodious songs, we may be tempted to imagine that the birds mimic just for the sheer beauty and challenge of it and because song is how they express their vitality and high spirits.

But Dalziell suspects there's something else at play.

One lyrebird in Dalziell's study area wins a lot of matings. She calls him Triple Blue and he sings especially beautifully, "a truly astounding free-flowing improvisation full of loud, flamboyant mimicry," she says. "For lyrebirds, the evidence is strong that mimicry in males is sexually selected." Belting out mimicry to females he can't see is a male's way of advertising himself, saying, "Look how clever I am, how accurate and versatile my mimicry is, how strong is my voice. I'm very intelligent (the accuracy of my mimicry shows that I have a good brain), and I can sing for hours a day." As Dalziell points out, the female alone sees to all the parental care—building the nest, brooding and raising the chick, defending her feeding territory. All she needs from a male is sperm. A choosy female measures a prospective mate by his spectacular plumage, his dancing ability, and his loud, complex, flamboyant song, including the versatility and accuracy of his mimicry and the carrying capacity of his calls. All of these traits may act as "honest" signals of a male's quality—his ability to learn, his strength and vigor, his genes, the conditions in which he was raised. "It's probably good for females to mate with males that are proficient vocal mimics so that their offspring are also competent at using mimicry to defend nests if they're female and to obtain mates if they're male," says Dalziell. A female will listen to various males for months and months before she settles on a mate. Only the best performers get to pass on their genes. Hence, a smart, high-performing bird species gets smarter, better at mimicking, at dancing, and at growing gorgeous tail feathers.

This is also true for other mimicking species, such as bowerbirds. For a satin bowerbird, spot-on mimicry appears supremely sexy. Mimicry seems to matter more to a female than either the beauty of a male's bower or the number of his ornaments, and males that accurately imitate many different sounds win more matings. It may not be just length and accuracy of mimicked songs and calls but also their difficulty to produce that points to a male's quality. In songbirds that don't mimic, it's the so-called sexy syllables—or high-performance trills that use both sides of the syrinx to produce simultaneous harmonious sounds—that females value most.

In superb lyrebirds, males get better at mimicking over time, so their accuracy appears to correlate with age and may signal a mate that's better at surviving than others.

Birds not only mimic to impress potential mates. Mounting research suggests they also appropriate calls and songs to deceive and manipulate others for their own profit. For instance, to steal a free lunch.

The now classic example is the fork-tailed drongo of southern Africa, a shiny little black bird with a split tail, red eyes, and a hooked beak. Tom Flower, a behavioral ecologist at the University of Cape Town, followed 64 drongos for a total of 847 hours and found that these "kleptoparasites" use mimicry as a deceptive practice to steal food from a variety of species. The drongo's repertoire of alarm calls includes 6 of its own and up to 45 mimicked calls from a range of birds and mammals, including the mongoose and the jackal. The drongo acts as an honest sentry for foraging neighbors, meerkats and southern pied babblers, perching high in the trees and warning of approaching danger with genuine alarm calls. But it also mimics specific alarm calls for larcenous purposes, falsely signaling an impending attack to startle a babbler or a meerkat into dropping its cricket or beetle and fleeing to cover so that the drongo can swoop in and snatch up the juicy morsel.

Here's what's really notable: If the babblers and meerkats get wise to the drongos' gambit and don't flee as readily, the drongos counter with a rare deceptive technique: They shift to different alarm calls from their extensive repertoire so their targets don't twig to their trickery. The birds even take their racket a step further, keeping track of whether they have duped an individual babbler and switching up the calls they use when attempting to steal again from that individual.

There are reports of blue jays mimicking not just red-tailed hawks, but raptors of all kinds, causing grackles and other birds to drop their food and flee, whereupon the jays seize the free meal. Even nestlings impersonate other young to glean more food. In some cuckoos and other brood parasites that lay their eggs in the nests of other birds, the parasitic chicks mimic the begging calls of their host's young to elicit meals. Individual common cuckoo chicks can imitate a whole brood of reed warblers, so its tiny parasitized parents step up their efforts to satisfy the oversize alien young.

Birds also pirate other animal calls and sounds to fool predators and escape being eaten. Several birds that nest in cavities, both nestlings and incubating adults, hiss like a snake when disturbed by predators. Eurasian wrynecks, little brown woodpeckers native to Africa, Europe, and Asia, writhe from side to side and hiss in their dark cavity nests to mimic a snake. When disturbed while incubating, Carolina chickadees will imitate the sounds of a copperhead. The northern flicker makes a buzz like a hive of bees to deter predatory squirrels.

But it's the tiny brown thornbill that most impresses in this regard. The bird is not much to look at, described as boringly brown and once dubbed by *Australian Birdlife* as the second dullest bird in the country after the mountain thornbill, which lacks even the modestly decorative speckled breast of the brown. But Brani Igic, who has studied this bird extensively, begs to differ. He says the thornbills may be visually inconspicuous, but they have a very big presence. Not only do their distinctive songs carry for great distances, but "they're just super charismatic," he says. He calls them the "original Angry Birds" because "they get extremely worked up, especially when they're protecting their nest. But they're also just really, really funny." Once, Igic was banding nestling thornbills, which involved temporarily plucking the baby birds from the nest. The parents returned while he was still in the midst of banding their young. "They were obviously very, very angry at us, screaming and scolding," he recalls. "But one of them had this huge caterpillar in its beak—probably half its size—and it was still trying to scream at us, with this flopping caterpillar in its beak. It was just so comical, I couldn't . . . I just had to stop what I was doing."

The archenemy of the brown thornbill is the pied currawong, a predator that feeds its own young with the nestlings and fledglings of small birds. "These birds are known as the Darth Vaders of the small bird world," says Igic. If you've ever seen one, you know why. They're big, black, scary-looking birds, thought to be highly intelligent, and they're voracious nest predators. A pair of currawongs will catch about 4.5 pounds of small nestlings, which comes out to about forty broods, to raise a single brood of its own. "And because they're so clever, they'll actually sit and watch for activity around a nest and then memorize nest locations," says Igic. "In fact, when we're studying nestlings, we worry that the currawongs will follow us to find the nests."

Pied currawongs are forty times the size of a thornbill, so the tiny bird has no chance of physically fighting off Darth Vader. But it does have a secret weapon: It cries wolf. Or, rather, hawk. Igic found that when a currawong attacks a nest, the thornbills send up multiple cries of alarm, mimicking a loud chorus of alarm calls by local species, creating the impression of impending attack by a predator that's scary even to the currawong—the brown goshawk. The little birds don't mimic the call of the big goshawk itself (which would be less effective—hawks don't call mid-hunt); instead, they imitate the flurry of aerial alarm calls normally spurred by its presence, including—and perhaps especially—the calls of those reliable sentinels, New Holland honeyeaters. In response, currawongs flee immediately or take a moment to scan the sky. Either way, the mimicked chorus distracts them long enough for the thornbill nestlings to scramble out of their little dome nests and hide in the thick vegetation around them.

Birds also steal songs to repel rivals or competitors. Regent honeyeaters mimic the calls of larger honeyeaters like wattlebirds and friarbirds, which may scare away rival birds that compete for the same food sources. When disturbed in their burrows, burrowing owls of the Americas rattle like an agitated rattlesnake to deter California ground squirrels or other competitors that might steal their burrows. Some years ago, a researcher proposed that northern mockingbirds mimic the vocalizations of lots of different species to defend their territory, creating the illusion of an area dense with competitors or predators. Female superb lyrebirds may do this, too, selectively mimicking the calls of hawks, perhaps to give the impression that the dangerous raptors are present, thereby decreasing the appeal of their territories to competitors.

A particularly sinister example of a bird using mimicry to gain a competitive edge is a forest-dwelling bird that lives on the island of Trinidad. When behavioral ecologist Julian Kapoor studied the little hermit hummingbird, he found deceptive mimicry taken to a whole new level.

One consequence of vocal imitation is that local dialects may arise, regional variations that allow neighbors to recognize one another as members of the neighborhood and to distinguish drifters from local territory holders (the "dear enemy" idea familiar from the red-billed quelea). This is the case with male little hermits. In breeding season, the birds perform courtship displays at leks, communal areas in the dense undergrowth of the forest. There, some five to fifty birds sing and display to win mates. The song of the little hermit is a high-pitched, squeaky chittering of five to eight syllables that lasts only a second but is then repeated for as long as an hour between foraging breaks. At their leks, the birds sing long and hard from about seven a.m. to dusk, up to twelve thousand songs in a single day. Their songs are highly variable, with lots of little local dialects, even within a single lek.

"It's crazy," says Kapoor. "An average lek is about the size of an average house lot. You can walk from one end to the other in a couple of minutes. And within a single lek, you'll get very distinct dialects. So it's like everyone in the living room is speaking French, and everyone in the kitchen is speaking Swahili. It's just totally bizarre that you would get dialects on that spatial scale."

To sort out why the little hermits might have these "microgeographic" dialects, Kapoor studied the bird's songs and their variations and how competition between males might shape them. During the breeding season, lekking males perch on a special slender twig and vigorously defend their territories by singing while wagging their little white-tipped tails. Most males occupy their singing perches and territories for less than a year, though a few lucky ones may persist as long as seven years. If you're a young male, it's hard to win your own territory. Young males seeking to join a lek use their gifts of imitation to impersonate existing territory holders. "First, they'll sit out on the edges of the lek listening," explains Kapoor. "Then they'll sneak around and eavesdrop on different individuals, focusing in on one local bird—let's call him 'Fred'—listening to him and learning his song. Then they'll go off into the forest and practice by themselves. They sound horrible at first, squeaking away, and for the first couple of weeks, you can't identify what they're trying to copy." But the young birds practice and practice, and over the course of a few weeks, they get much better. "By the time a young male is ready to come back and take his place on the lek," says Kapoor, "he sounds *exactly* like Fred—or at least, one of his close neighbors—so I know exactly where in the lek he's intending to go."

And here's the creepy part: Kapoor suspects that a newcomer "mimic male" may attempt to usurp an established male's territory by attacking and killing him and then taking his place—if he himself is not killed in the process. This is still just a hypothesis, yet to be observed, but Kapoor has seen brutal fights that lasted a week or more, with the birds trying to stab each other with specialized sharp tips on their bills. If a young mimic male is successful, "thereafter, he would likely be treated by the other birds in the lek as if he were actually Fred, their local neighbor, and tolerated as the rightful owner of his territory," he says. "The other territory holders wouldn't know the difference because his mimicry is sufficiently well practiced to fool them."

There's no evidence to suggest that lyrebirds attempt to fool other species for their own gain, but like the little hermit, they may attempt to fool their own kind. Dalziell and her colleague Justin Welbergen recently made a stunning discovery while watching pairs of lyrebirds mating. Two of the matings they witnessed in person and several others they caught on video.

"Not all of a male lyrebird's mimicry is equal," says Dalziell. "It has different phases and functions during display," including—it turns out— one singular finale with a duplicitous purpose.

"It's very stark," she says. "Most of the male lyrebird's mimicry while it's in breeding mode occurs during perch singing, called recital song because the birds just sit there and perform," she explains. That's followed by the full song-and-dance display on the mound.

But then, at the end of this display, the birds sometimes do something strange. While the female is crouching on the mound, a male will switch very suddenly from mimicking songs to mimicking alarm calls. "And not just any alarm calls," says Dalziell, "only those uttered specifically by ground-dwelling birds—brown thornbills, white-browed scrubwrens, yellow-throated scrubwrens, eastern yellow robins, eastern whipbirds— in response to a major ground threat, such as a snake or a cat or other threatening mammalian predator." He imitates not just a single bird but many different birds producing alarm calls all at once, just as the brown thornbill does. "He'll run the elements together to create an illusion of multiple birds alarm-calling, some near, some farther away," she explains. "What he's doing is mimicking the sort of live mobbing group chorus you would actually get around a threatening snake."

Why in the world would a male lyrebird do this in the middle of mating, with no snake in sight?

Dalziell and Welbergen have a theory. The longer a male copulates, the better his chances of successful fertilization. To keep the female in place on the mound, he uses his masterful mimicry to deceive her, frightening her with a false chorus of alarms that may induce a "freeze" response or a reluctance to leave the male's display mound. Some male moths also do this during copulation, mimicking the ultrasonic echolocation calls made by moth-eating bats to immobilize the female moth and further their chances of fertilization. In both bird and moth, the male is catching the female in a "sensory trap," mimicking an alarm call and exploiting her fear response. She can't afford to evolve resistance to such male trickery because the consequences of ignoring a mobbing chorus could be fatal. In this way, the male lyrebird uses his gifts of mimicry for vocal deception, misleading his mate, manipulating her biological response—lying to her—so she'll crouch in terror while he finishes his business.

"That's quite a different story from the one we grew up on, of mimicry as an honest signal of fitness and mental agility," says Dalziell—and quite at odds with the image of the lyrebird as the mother of all songbirds.

Work

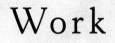

Chapter Four

THE SCENT OF SUSTENANCE

S ome years ago, I spent freezing twilight hours in the northern woods of Hokkaido, Japan, waiting for a hungry bird to descend from the darkening forest and feed from freshly killed fish offered up on a table by researchers. At first the woods were silent. Then we heard the booming *BO-BOHHH* of a male, and a few seconds later, the quieter *bohh* of a female in response: the simple duet of the endangered Blakiston's fish owl, the biggest owl on the planet, up to ten pounds and nearly three feet high, with a six-foot wingspan—think great horned owl with triple the mass—master of the dusk hunt, and a deeply secretive species.

The fish owl does not usually get its dinner served up on a buffet like this but must work for it, sit patiently on a riverbank for long hours, watching the icy riffles with wild gold eyes. When it spots the little back of a fish or the fin poking from the shallows, it jumps in feetfirst and snatches its prey with talons sharp enough to open a can. In autumn it hunts migrating salmon so big it can barely carry them, says Jonathan Slaght, who has studied the fish owl for fifteen years and manages research projects on the species for the Wildlife Conservation Society. One owl was seen trying to lift a huge fish struggling beneath his grip. It looked to the riverbank, spotted a tree root, and grabbed it with one talon, using the root as leverage to haul itself and its fish onto the bank.

What is true for us is true for birds, too: They are what they eat. A healthy diet means beautiful bright feathers attractive to mates, the development of fine wiring for song learning and performance, a better memory for where to find food. No wonder passerine parents feed their young large numbers of spiders—a rich source of the amino acid taurine, thought to be essential for normal brain growth and development—and Blakiston's fish owls feed their nestlings a steady stream of fish, frogs, and the occasional lamprey.

Like humans, birds will go to great lengths to get a good meal. They have figured out how to dig up invisible food, crack open its armor, discard its poisonous or unpalatable parts, and fish it out of hidden places with specially crafted tools. They've learned how to process impossible amounts of spatial information to remember where food is, manipulate their prey, even detect food through senses we thought they didn't have.

Some birds take on extremely challenging prey. Consider the short-toed snake-eagle, a raptor of southern Africa with a name like a hybrid from Greek mythology. When the snake-eagle spots its reptilian prey from high in the sky, it dives down, seizing the snake with its talons and lifting off again with the thrashing, striking serpent in its clutches. While still in flight, it crushes or rips off the head, then swallows the beast whole. With equal delicacy, its cousin the brown snake-eagle consumes venomous snakes such as black mambas and cobras up to nine feet long.

I've never seen a snake-eagle take its writhing prey, but I have seen a great blue heron wrestle with a large water snake for close to thirty minutes—and win.

A video camera on the nest of a Blakiston's fish owl, placed by Slaght's colleague Sergei Surmach, captured footage of an adult bringing a Pacific lamprey to its fat little nestling in the cavity of a giant tree. The film shows the chick wandering the perimeter of the nest all night, shrieking for food. When the adult shows up, it has a live lamprey two feet long flip-flopping around in its bill. "The nestling shuffles over eagerly and takes this thing in its mouth," says Slaght. The adult flies away. And then the next few minutes of the video is the nestling trying to negotiate the thrashing lamprey in its beak. "The lamprey's tail is flapping wildly back and forth, hitting the nestling in the face," says Slaght, "while the bird is using this ratcheting motion to try to choke the thing down. Finally, the young owl swallows it. There's a lag time while he's quietly sitting there. Then the adult again lands on the edge of the nest—with another live lamprey—and the nestling takes this little step back."

The loggerhead shrike, best known for impaling its prey on a twig or piece of barbed wire for future consumption, tackles its bigger victims— mice, frogs, birds—with a gangster's way of killing. One shrike was spotted striking a cardinal from behind and holding on to it during a noisy struggle that lasted less than a minute. Then the shrike flew off carrying its cumbersome load low to the ground, dragging it for short distances. A study by Diego Sustaita and his colleagues in 2018 with the somewhat macabre title "Come on baby, let's do the twist" revealed the details of the shrike's murderous attack. These birds repeatedly bite the head or neck of their victims with their sharp beaks, then latch onto the nape and vigorously shake it with rapid, snappy head rolls until they break the neck and damage the spinal cord, inducing paralysis. Other predators shake their prey in this way, roadrunners, lizards, snakes, and some mammals, but only shrikes and crocodiles "do the twist."

A population of great tits in northeastern Hungary, short of the usual food resources, has taken to eating hibernating bats roosting in the huge Istállós-kői cave. The tits swoop into the cave and emerge ten or fifteen minutes later, bat in beak, often settling in a nearby tree to feed on the little mammal headfirst. That birds as small as tits can be such fierce hunters and make such a drastic shift in their prey is a surprise. But necessity is the mother of invention.

Some birds have learned to dissect their prey, allowing them to eat animals that would otherwise be highly poisonous. I once watched a pied butcherbird eat a deadly cane toad in the suburbs of Brisbane. Think about what's involved in this. The new-world toad was introduced to

Australia in 1935 and spread quickly from Cape York Peninsula to Sydney and west beyond Darwin, bringing death to predators unfamiliar with its venom. Over just a few decades, pied butcherbirds, along with a few other clever species—Torresian crows, black kites, pied currawongs, white ibises—discovered the trick of avoiding the toad's poisonous parts, the glands on the back of its head and the skin itself. Otherwise the toad is well worth eating. Their technique: flipping over the toad and eating from the belly side inward. Ibises consume only the toad's legs. Black kites eat only the tongue.

Kea parrots of New Zealand eat more than a hundred different plant species, some of which are highly poisonous, with roots, stems, leaves, and seeds all containing toxins. Only the fruit pulp is edible, and kea have learned to root it out.

Before consuming autumn gum moth caterpillars, a crested shrike-tit will remove the digestive tract, which contains a poisonous oil from the eucalypts the caterpillars feed on.

Vultures will leave aside the scent glands of dead skunks.

Vultures: They're often considered even more ghoulish than shrikes, but for different reasons.

I live down the road and across the railroad tracks from Hog Waller, the site of an old livestock auction house. It sits in a low-lying area where heavy rain occasionally makes a muddy mess of the stockyards, where hogs and pigs may wallow (or "waller," as it's pronounced by locals). The livestock market had its heyday in the 1940s and 1950s. When I moved here in the 1990s, it was a sleepy place, with only the occasional cattle cart rattling past with a load of wide-eyed cows. But you knew when the place was active by the volt of vultures across the road hunching awkwardly on the rooftops, sunken heads rounding into curved shoulders. Today they still jostle for space near the site, like a Supreme Court made of Richard Nixons. A mix of black and turkey vultures, both with that undertaker's shading, they're a lugubrious sight, made more so by knowledge of their gross habits: a penchant for peeing on their own legs to cool themselves on hot days, vomiting their entire stomach contents when they're under attack, and most of all, shoving those homely, featherless heads deep into the gore of a carcass to feed on its soft tissues, its eyes, mouth, and anus. They let me come near. Then with a slow, reluctant nonchalance, they lift up, flap, settle heavily on the next roof over.

The knock on vultures is that they're dark plunderers of rotting meat and purveyors of disease, connoisseurs of the bad heap, the rank dish of roadkill, the bones, meat, skin of the dead. That's what I was taught growing up. And that's what Charles Darwin thought when he spotted a turkey vulture from the deck of the *Beagle* in 1835. He called it a "disgusting bird" whose bald head was "formed to wallow in putridity."

But as most of us now know, turkey vultures have gotten a bad rap. The truth is, those naked heads relieved of adornment are extremely hygienic—gory stuff just doesn't stick. And it's a lie that vultures particularly relish a ripe, rotting mess. They actually prefer fresher carrion. As such, they perform a vital, and vastly underrated, service to the environment: the quick, competent cleanup and recycling of dead creatures.

Vultures are nature's sanitary workers. Because they feed in groups and eat rapidly—each bird downing more than two pounds of meat a minute—they can rapidly consume whole carcasses. Their guts are acidic enough to destroy the agents of disease, such as cholera and anthrax, so there's little risk of spreading contamination from an infected carcass. That's not the case with more leisurely mammalian carrion eaters such as rats or dogs or coyotes.

What happens when vultures vanish, people learned the hard way in India and Pakistan more than a decade ago. There, a mass die-off of old-world vultures caused by an arthritis drug in the flesh of dead cattle led to an eruption of rabies. Dogs took over feeding on the carcasses, and the canine population exploded, along with the spread of the deadly disease.

While vultures may be clumsy on foot, they're aloof and beautiful in flight, big as a bald eagle, but with upturned wings and flight feathers almost silvery from below. They're masters at catching rising thermals and updrafts even in calm winds, spiraling upward rapidly to impressive heights, where they appear to float motionless, tipping from side to side in their constant search for carrion.

People once thought this high circling flight was a sure sign that turkey vultures spotted their food by sight. We'll see just how wrong this was. But at the time, it was a reasonable assumption. In the nineteenth century no less a figure than the great naturalist and artist John James Audubon had proclaimed—and proven, he said—that turkey vultures locate food by vision alone, spying it from great heights.

Indeed, for a long time, most birds were considered merely "a wing guided by an eye." Foraging was a matter of instinct and visual searching out of sustenance—whether in the form of flitting insects, skittering rodents, fresh roadkill, or tasty and nutritious fruits, nuts, and berries.

This bias toward vision is hardly surprising. We humans are eye-minded creatures, and when it comes to visual acuity, we're near the top of the animal chart. It makes sense that for a long time, studies of food searching in birds focused on sight. The world of birds, we thought, was a world like ours, of light, color, and movement, the twitch of a rodent or the whirl of flies circling a carcass.

Some birds do have extremely keen powers of vision. Sharp as the human eye is, it doesn't hold a candle to a wedge-tailed eagle's, which can see three to four times farther than we can. An eagle's retina is more densely layered with cones than ours is, making its visual acuity, the ability to perceive detail and contrast, far better than ours. In the center of its eye is a specialized deep fovea—a little convex pit coated with cones— that may give its eyes extra magnification in the center of its field of view, like a telephoto lens, the better to spot that field mouse from hundreds of feet.

Birds top us in color vision, too. They see hues beyond our imagining. Humans have three types of color-receptive cones in our retinas, blue, green, and red. Birds have a fourth color cone that is sensitive to ultraviolet wavelengths. We are thus "trichromatic," and most diurnal birds are "tetrachromatic." With their extra UV cone, birds can distinguish shades of color we can't tell apart, allowing them to spot prey well camouflaged against the uniform background of a grassy field or leafy forest floor, and to detect things invisible to us—like the trail of urine left by a vole.

But it goes beyond this. Birds see a massive spectrum of color our brains are simply incapable of processing. "It's not just that they can see wavelengths of colors in a part of the spectrum we can't see," says Mary Caswell Stoddard, an assistant professor of ecology and evolutionary biology at Princeton University, who studies avian color vision. "It's that ultraviolet light is a fundamental part of many of the colors they perceive. They're experiencing another whole dimension of color—all the colors we can see, with varied amounts of UV mixed in. So it's not simply human vision plus some purplish UV colors. It's a complete reimagining of the color experience."

While there's no way to know what a color actually looks like to birds, Stoddard can measure with a spectrophotometer the wavelengths of light reflected by an object or a surface. This tells her whether or not a color is reflecting ultraviolet light. Then she can estimate its appearance in a way that's relevant to bird vision using a computer program called Tetra-ColorSpace that she developed with ornithologist Richard Prum.

Take, for example, Stoddard's favorite bird, the painted bunting, which she grew to love on family visits to Florida. "The bunting's back is a spectacular shiny green color," she says. "When you put the spectrophotometer on the green back of the bunting, there's a green peak, showing that green wavelengths are being reflected, which is what we're seeing. But there's also a huge ultraviolet peak that we're completely missing. To a bird, the green back of the painted bunting looks UV green, which isn't green at all but a totally different color that we can't even fathom."

Stoddard explains that it's kind of like the distinction between black-and-white and color television. "If you wanted to explain to someone who has only watched black-and-white television what color TV is like, it would be hard," she says. "Well, the distance from our color vision to a bird's color experience is probably a little like that, a quantum leap."

In 2019, biologist Dan-Eric Nilsson and his colleagues at Lund University in Sweden released photos from a camera designed to re-create how birds see color in their surroundings. With the help of special filters, the camera simulates the full visual spectrum of birds. One major discovery: To birds, the dense foliage of a rainforest likely looks not like the uniform, largely flat mass of green we see, but a detailed three-dimensional world of highly contrasting individual leaves. I have tried to see this way, to pick out distinct leaves amid the wall of green, but contrasts in green are hard for humans to distinguish—which is what makes the green catbird in the rainforest so impossible to spot. UV light amplifies the contrast between the tops of leaf surfaces and their undersides, so the three-dimensional structure—the position and orientation—of the leaves pop out. This makes it easier for birds to navigate through complex leafy environments and to find food there.

Color, beauty, reality truly do lie in the eye of the beholder.

The Blakiston's fish owl in Hokkaido may look for the ripple of fish in a shallow stream, but most owls listen for their food.

In 2018, I spent an hour in an aviary with "Percy," a great gray (Lapland) owl, a species that listens for voles and other rodents tunneling through loose soil and can hear them at a distance of hundreds of feet, even if they're beneath two feet of snow. A zookeeper at the open-air museum of Skansen in Stockholm, Sweden, lured Percy toward my arm with a bowlful of a dozen defrosted mice. Being close to a bird as magnificent as this, even in the aviary of a zoo, took my breath away. In keeping with its name, the bird is enormous, three feet high, with a five-foot wingspan, but without the heft of a Blakiston's. Its bulk is nearly all feather, and its flight is velvety silent despite the bird's size. It has a huge head and a broad, flat face that swiveled toward me like a lighthouse beam to stare with bright yellow eyes. The facial disc—the circle of feathers around each eye—acts as a pinna or dish antenna, directing sound to the owl's colossal and complex ear openings, which are asymmetrical; the left ear opening is farther upward than the right. This asymmetry allows the owl to pinpoint the direction of a sound from its perch. And also, to locate prey in total darkness, as a barn owl does, with an error of less than one degree. The hunting owl listens intently, turning its head from side to side. Once it detects the rustlings of a rodent, it dives down headfirst, at the last instant swinging its feet forward beneath its chin to snag its prey.

If you think about where owls are hunting, the natural conditions they face, the Blakiston's reliance on vision and the great gray owl's reliance on hearing make sense, says Jonathan Slaght. Great gray owls hunt where it's flat, with snow and quiet. Fish owls regularly hunt in narrow river valleys, with shallow, fast-flowing, and rocky waterways. "If a fish owl had hypersensitive hearing and hunted in these waterways, it would go crazy," he says. "It's so loud that trying to isolate the sound of prey in that type of habitat would be impossible. The fish owl does have a facial disk, but it's nowhere near as pronounced as that of a great gray owl or a barn owl, which can catch things in 100 percent darkness because of their heightened hearing ability."

For a long time, scientists thought birds smaller than owls were poorly adapted for finding prey through hearing because their small heads don't create enough of a "sound shadow." Their ears were too close together to generate that little interaural time difference that helps an owl localize sounds. But studies of the past few decades show otherwise. The Australian magpie can find the buried larvae of scarab beetles just by detecting the faint scratching sounds made by the burrowing of the larvae. When vision is limited, American robins search for earthworms by sound. Watch them in a field or on a lawn: They run a few steps, then lunge and drive their bills deep into the soil, more often than not snagging an earthworm. Scientists have recorded capture rates as high as twenty earthworms per hour. Mask all sound with white noise, as the researchers did, and the robins are not nearly as successful.

A handful of birds use the sound of their own voices to move around in darkness, including oilbirds of tropical South America. These birds were first discovered by the European scientific world in 1799, when

Prussian naturalist Alexander von Humboldt followed members of the local indigenous tribe to a cave called Caripe, the "mine of fat," in northeastern Venezuela. The expedition followed a small river to the mouth of the cave. As the group moved into the darkness, hoarse screams, piercing shrieks and snarls, and grotesque retching sounds reverberated against the rocky vault and echoed in its depths. The cave was full of several thousand birds known as *diablotin*, "little devil," and locally as *guácharo*, "one who cries and laments." When Humboldt's guides raised their torches, the light revealed the source of the horrid sounds: thousands of funnel-shaped nests fifty to sixty feet above them. The men shot their guns upward in the dark, bringing down two birds, which Humboldt later examined. "The guácharo is about the size of our chickens," he wrote, "with the mouth of our goatsuckers and the gait of vultures, with silky stiff hair around their curved beaks."

The oilbird is the only species in its own genus. Its common name comes from the young birds that become quite plump before they leave the nest, weighing half again as much as their parents. Their oil is clear and odorless, wrote Humboldt, "and so pure it lasts for a year without going rancid." Indigenous people had known about the bird for centuries and harvested its fat to flavor food and fuel torches, but they took only the birds at the front of the cave, so the population survived.

The oilbird is the only nocturnal fruit-eating bird in the world. The huge flocks of several thousand birds spend their days roosting and screeching in the blackness of the cave. Every night they set out in silence to forage in the surrounding forests for fruits of palm and laurel trees, which they swallow whole, regurgitating the seeds.

Imagine living so deep within a cave that no daylight penetrates, and not once in your life experiencing any illumination brighter than the moon. The oilbird has developed tools to negotiate its dark world: It has special long bristles around its beak, which it uses for tactile sensing, the better to probe fruit and other necessities through touch. It also has the most-light-sensitive eyes of any vertebrate on earth, with a higher density of rods in its retina, some one million per square millimeter, which allow it to see at night while foraging in moonlight. But in the total darkness of the deep cave, where the bird roosts and nests, it needs a guidance system beyond touch or vision.

These days we take it for granted that some animals use reflected sounds they make themselves to detect objects in their environment. But when Donald Griffin and fellow student Robert Galambos first proposed this notion at a scientific meeting in the 1940s, they were practically laughed off the stage. Griffin later wrote that one distinguished physiologist was so shocked by their presentation on bats "that he seized Bob by the shoulders and shook him while expostulating, 'You can't really mean that!'"

It took years for scientists to fully accept the idea of echolocation, as Griffin termed it. Decades later, intrigued by reports of the nocturnal oilbird, Griffin tested the strange creatures and revealed a surprising fact: Plug the ears of the birds in a pitch-dark room, and they collide with the walls; unplug their ears, and they easily avoid them.

Unlike bats, which emit sounds at ultrasonic frequencies too high for us to hear, oilbirds emit audible clicks, which bounce off objects and provide an auditory map of their surroundings in the dark. The birds make the clicks with their syrinx, opening and closing their vocal folds five times per second, closely matching their rate of five wingbeats per second when they fly. Danish biologist Signe Brinkløv has observed them clicking while entering and exiting caves and adjusting the number of clicks to the overall lighting conditions. "By sending out longer click bursts when there is less moonlight available, generating a signal with more energy, they're able to reflect bigger echoes for a given distance," she hypothesizes, "which increases the range of their echolocation system when lighting conditions are unfavorable."

Not only do oilbirds have touch-sensitive whiskers, light-sensitive eyes, and an ability to echolocate, they also have big olfactory organs, suggesting that smell may play an important role in their foraging.

Birds find food by looking, listening, even echolocating. But smelling? Well into the middle of the twentieth century, people thought birds could barely smell at all, much less forage with their noses. Even when anatomical studies showed that some birds, such as turkey vultures and oilbirds, had very large olfactory bulbs—the part of the brain responsible for processing odors—there was still great skepticism about whether the bulbs actually gave birds the functional ability to detect odors. Contradictory observations and studies, many of them poorly designed, contributed to the confusion and created the general impression that birds simply didn't use their olfactory equipment.

When Gabrielle Nevitt decided to probe the possibility that certain seabirds might use smell to pinpoint their prey in the immense, featureless ocean, she faced the same skepticism borne by Griffin and Galambos. Many scientists, if not most, considered the idea that birds could smell absurd. In fact, this dogma still prevails, says Nevitt. "Look at any ornithology textbook, and it's likely to say birds lack a good sense of smell." Now head of the sensory ecology lab at the University of California, Davis, Nevitt has twenty-five years' worth of brilliant studies unraveling this canard.

Nevitt can fault John James Audubon for the skepticism she faced. Almost two centuries ago, the grand old man of bird illustration planted the seed of doubt about whether birds can smell. At a public talk in 1826, he recommended that people "abandon the deeply-rooted notion" that turkey vultures discover their prey by smell. To back his argument that vultures do not scavenge by nose, Audubon had "assiduously engaged in a series of experiments," as he put it.

"Incredibly interesting" is what Nevitt calls these investigations, with more than a whiff of irony.

To start, Audubon stuffed a deerskin with grass. This he placed on its back in the middle of an open field, with legs up and apart, "as if the animal was dead and putrid," though in fact it was odorless. To his great satisfaction, "A Vulture, coursing round the field tolerably high, espied the skin, sailed directly towards it, and alighted within a few yards of it," he said. "Then approaching the eyes, that were here solid globes of hard, dried, and painted clay, attacked first one and then the other." From this demonstration, Audubon concluded that the bird's ocular powers bested its sense of smell.

For his second experiment, Audubon hauled a dead hog into a ravine, covered it with briars and cane stalks, and left it to decay for two days in the heat of July, "when a dead body becomes putrid and extremely fetid in a short time," he wrote. This time, the vultures circled the ravine, but none neared the reeking carcass. Several dogs did, however, and fed on it, though the smell was "insufferable." To the vultures, too, apparently. As it turns out, vultures don't relish putrefied meat; they prefer a fresh kill. Audubon's experiment was a little like serving your cocktail guests fetid oysters and surmising they don't have a taste for seafood.

The results were "fully conclusive," he said. "The power of smelling in these birds has been grossly exaggerated."

Later observers following Audubon's lead went out of their way to explain how turkey vultures might find hidden food by any means *other* than smell. By watching or listening for the buzzing of flies, for instance, or by observing the movements of carrion-eating mice and ground squirrels scurrying to and from a concealed carcass, or domestic dogs sniffing it out with their keen noses. Vultures perhaps have such big olfactory bulbs not to spot prey, they theorized, but to detect the direction and quality of air currents for soaring.

Fortunately, Betsy Bang came along to crush Audubon's theory. Bang, a scientific illustrator at Johns Hopkins University in the late 1950s, dissected and sketched the nasal cavities of various bird species to illustrate her veterinarian husband's articles on respiratory disease in birds. She eventually created an entire atlas of olfactory structures in bird brains, which graphically pointed to their notable size and prominence in

some species. As Gabrielle Nevitt tells it, "Bang was a budding naturalist and birder herself, and she thought, 'Oh my goodness, birds have a
sense of smell!' And when she got into the literature, she found that 'Oh
my goodness, a lot of textbooks say that they don't!'"

"It seems curious that the large olfactory organs of certain species
have so often been pointed out by anatomists," Bang wrote, "yet most
olfactory learning studies have been done on feebly equipped birds such
as pigeons and have tended to keep alive in textbooks the idea that the
chemical sense in birds is minimal or lacking."

Bang wrote a series of papers detailing the anatomical evidence for
impressive smell abilities in a variety of birds, focusing on the brain and
peripheral olfactory system. One illustration showed the massive turbinate structures in the olfactory system of the turkey vulture, which contrast sharply with the smaller turbinate structures of black vultures.
"Not only did Betsy suggest that some birds had a fine sense of smell,"
says Nevitt, "but she also introduced this idea of comparing the sense of
smell between closely related species. From one artist to another, Betsy
declared, 'Audubon, you're wrong: Birds *do* have a sense of smell.'"

A decade later, ornithologist Kenneth Stager bolstered Bang's ideas,
demonstrating that turkey vultures do indeed sniff out their carrion and
that they prefer fresh carcasses to putrid ones. In one experiment, he concealed the freshly skinned carcass of a badger wrapped in newspaper in
the heart of a dense creosote bush. A lone turkey vulture showed up, circled the area first upwind, then downwind, flying low over the bush.
"After a short inspection of the area," wrote Stager, "the vulture walked
directly up-wind toward the bait, and with its beak, pulled the paper-
wrapped carcass from the center of the creosote bush."

This was just the beginning. Stager also identified the specific scent
that drew vultures to carrion. The discovery arose from a serendipitous
conversation with field engineers at an oil company, who had been aware
for some time that turkey vultures have knowing noses, and used the
birds to locate leaks in natural gas lines. The engineers had figured out

that if they introduced ethyl mercaptan into the line, they could locate leaks by the concentrations of turkey vultures circling above the line or sitting on the ground next to it. The same sulfurous chemical added to natural gas so the human nose can detect a leak, it turns out, is also released by an animal shortly after it dies. Thanks to Stager's experiments, we now know that turkey vultures are highly attracted to the scent and can use it to find a target as tiny as a dead vole buried in leaves on the forest floor beneath a thick canopy of trees.

Just what makes the nose of a turkey vulture so supersensitive came to light in 2017 when scientists revisited Bang's idea and dissected the brains of both black and turkey vultures to compare them. Gary Graves, an ornithologist at the Smithsonian Institution, had heard about a USDA culling operation of black and turkey vultures in Nashville, Tennessee, and decided to make use of the brains of the freshly killed birds. Not only does the turkey vulture have an olfactory bulb four times the size of the black vulture's, the study showed; it also has twice as many mitral cells, specialized neurons that transfer information from smell receptors to the parts of the brain that interpret it, allowing the brain to distinguish between many different kinds of scents and to link a scent with an object. With three times as many mitral cells as a rabbit or a rat, a turkey vulture can detect fragrant molecules wafting into the air from decaying flesh in concentrations as tiny as a few parts per billion.

What exactly is the bird sniffing? Scientists aren't sure. It could be a single scent like ethyl mercaptan or a whole complex cocktail of hundreds of volatile compounds that make up the odor of death. Each type of decaying tissue—muscle, skin, fat, etc.—releases its own perfume. What we do know is that these birds can track plumes of odor from a carcass while gliding high in the air and then home in on it by circling, moving down the odor plume until they find its source.

The turkey vulture is one of the bloodhounds of the bird world, and because of its superior ability to find food beneath tree canopies by scent alone, it's the most successful species of vulture, with around eighteen million birds picking clean the carcasses of the world. But it's not the only bird with a keen sense of smell.

Beginning in the 1960s, physiologist Bernice Wenzel decided to explore whether birds of different species had working noses. "Bernice was a powerhouse in the field," says Nevitt. "She was well known for using the same state-of-the-art techniques to study avian olfaction that were used to study mammals at that time. She tested a range of taxa with the aim of demonstrating that the noses of birds were not just anatomically developed but also functional. Her studies very much complemented Betsy Bang's anatomical studies."

Wenzel found that every bird she tested could detect odors—pigeons, quail, hummingbirds, robins, even house finches and starlings. It turns out that European starlings use odor to discriminate between plant species during the nesting season, selecting fresh green plants that are rich in volatile compounds such as milfoil to line their nest holes. The aromatic herbs serve as a fumigant to protect nestlings from the load of parasites and pathogens that increase with repeated nest use. Interesting to note: The birds can distinguish between milfoil and other plant odors only seasonally, during courtship and nest building. The honeyguide uses scent to search out beehives, and the Oriental honey buzzard, a bird of prey in Taiwan, uses smell to track down nutritious balls of pollen-containing dough that Taiwanese beekeepers provide as supplementary food for their bees. The kiwi of New Zealand, the only bird with a nostril at the tip of its beak rather than at the base, uses its nose to unearth worms, invertebrates, and seeds. Puffins have such a fine sense of smell that they can find their colony by scent alone from a distance of nearly five hundred miles. House finches can detect predators by smell. Even ducks have a well-developed olfactory system. In fact, birds of all stripes use their sense of smell to navigate, locate burrows and nests, detect chemical signals during courtship, pick mates, avoid predators, and search out food. But as Gabrielle Nevitt has discovered, turkey vultures as sovereigns of olfactory foraging have only one set of real rivals in the bird world.

Nevitt is enthralled by smell. We met at a gathering of two thousand ornithologists in Vancouver, Canada, many of whom still held old textbook beliefs about smell in birds. When Nevitt took the stage, their old ways of thinking flew out the window. Nevitt started her career studying salmon at the University of Washington in Seattle, specifically their ability to smell their way home, tracing odor back to the spawning streams where they were born. But birds have always been her first love.

"A lot of my colleagues in ornithology seem to have been drawn to the field because they grew up watching wild birds, but for me it was *smelling* tame ones," she says. "My mother loved birds, and we had many very tame birds to play with. We had a parrot and a mynah bird and lots of chickens, both inside and outside the house." It was these pet birds, not books, that taught Nevitt about bird senses and behavior. "Our parrot was particular about what type of toast he ate and whether we buttered it with the real thing or margarine. I carried a pet bantam rooster around with me like a lot of little girls carry a doll, so I remember the sweet smell at the base of his comb and that the heckle feathers on his neck smelled like a plum tree, or grass, or a eucalyptus tree at times. I didn't know anything about chemistry, but I thought his feathers carried the scent of his surroundings, and I could always tell where he had been hanging out. Our mynah smelled dusty, except on his earlobe. His breath smelled like fruit or dog food, depending on what he had been eating. Being up close and personal with a variety of pet birds gave me particular insights that I wouldn't have had otherwise."

These days, Nevitt keeps several birds, including emus and peacocks. "A few years ago, we moved out to the country, in part so we could have emus," she says. "I wanted to study them because they are basal"—near the base of the bird evolutionary tree—"and have large olfactory bulbs, and there was no place to house them on campus. We got them as chicks, but ironically, they turned into much-loved pets instead of scientific subjects." Since then she has acquired turkeys, peafowl, and various chickens. "We have geese as watchdogs," she says. "They also have a good sense of smell."

Nevitt's switch from studying salmon to studying seabirds—known as the "fish of the air" because they wander the sea and go to land only to breed—seemed like a natural step. But it was met with stinging skepticism. "I was stunned by how deeply ridiculed I was," she says. "People told me I was nuts, that I'd never get my research funded." Nevertheless, she persisted, focusing her genius on tube-nosed seabirds—albatrosses, petrels, fulmars, and shearwaters. These birds are famous for the strong odor of musk that permeates their plumage. They're called tube-nosed because of the elongated tubes or nostrils on their thick upper beak, which they use to eliminate salt from the seawater they drink and, as Nevitt discovered, for another equally vital purpose.

Scientists knew that tube-nosed seabirds had a lot of fancy olfactory equipment. Detailed comparative bird brain studies like Bernice Wenzel's revealed that olfactory tissue takes up about 37 percent of a seabird's brain, versus about 3 percent in the brain of a typical songbird. Moreover, the olfactory bulbs of some seabirds have twice as many mitral cells as rats and six times as many as mice.

Until Nevitt came along, however, no one had fully explored how seabirds might use this elaborate nose gear to track their prey over vast ocean expanses. These birds roam the sea, soaring over extreme distances, to search for small shifting patches of prey, krill, fish, and dead squid floating on the ocean surface—for albatrosses, thousands of kilometers in a single foraging trip. "Many of these species routinely travel in the dark or under conditions where visibility is limited by fog or extreme cloud cover," says Nevitt. "Their survival depends on finding the proverbial needle in a haystack on a daily basis."

How do they do it?

By sniffing out scented compounds—one chemical in particular, dimethyl sulfide, or DMS, generated when krill devour phytoplankton. Over the past two decades, Nevitt has parsed the impressive ability of seabirds to detect tiny amounts of the chemical.

To our noses, DMS is that sulfury, briny smell of the seashore or oysters on the half shell. To seabirds, it's the scent of sustenance. "Birds tend to be attracted not to prey scents per se," she says, "but rather to odors such as DMS that are released during feeding interactions"—a euphemistic term for the ravaging of prey by predators. In other words, says Nevitt, "predators tend to be messy eaters, and tube-nosed seabirds have adapted to pay attention to who is eating whom."

A sensitivity to DMS also explains why seabirds consume plastics and other trash. Plastic debris emits the chemical scent, making trash smell like food and creating a kind of olfactory booby trap for seabirds.

You would think that a seabird sniffing its way to prey might follow a scent gradient, moving toward the concentrated stink. But airflow over oceans is not neat and orderly. It's turbulent and choppy. The odor plumes are more like cigarette smoke, coiling and drifting over the ocean surface. To intersect the plumes, seabirds fly in a zigzag, upwind pattern typical of bird dogs, fish, and other olfactory-tracking creatures.

Storm petrels are especially good at the strategy, detecting the chemical at great distances and tracking back and forth with the wind currents, continually sniffing the air until they zero in on their target. Other seabirds that rely more heavily on visual cues follow the petrels to food. The petrels themselves have a weasel-like smell, which may help them detect parents or mates. Ornithologist Edward Howe Forbush called them "peculiar, eerie birds" with habits so strange "that from time immemorial sailors have had a superstition that these small fowls are the precursors of storms and wrecks." The birds grow up in dark burrows, where smells are the dominant sensory experience, so they're more tuned to chemical cues. Nevitt has found that chicks of the burrow-nesting petrels can learn odors linked with prey while still in the nest. The baby birds are capable of detecting DMS and ammonia (a urinary by-product of most marine organisms) at minute concentrations—a million times lower than any odor sensitivity reported for birds in the past. Present a Leach's storm petrel chick with a vanishingly small amount of one of these odors, and the tiny bird will sweep its head in a broad arc around its body while making coughing noises and rapid biting movements close to the source of odor.

When Nevitt first looked at maps of DMS plumes across the ocean, she noticed something telling. The plumes seemed to overlay ocean features such as fronts and seamounts and other upwelling zones, where phytoplankton tend to accumulate, along with the krill eaten by seabirds. To a seabird, the ocean is nothing like the featureless expanse of water we see. It's an elaborate landscape of eddying odor plumes that reflect the oceanographic features and physical processes where phytoplankton predictably amass. "We speculate that the birds build up a map of this olfactory landscape over time from experience," says Nevitt, "and use it to guide them to likely areas for prey."

This is the sort of news that turns the kaleidoscope for a new worldview: of seabirds as feathered hounds, yes, but also of the planet itself, the air above the ocean as an invisible landscape full of subtle, swirling features sensed by creatures far better attuned than we.

And here's a postscript: We've known for some time that colonies of cliff swallows, like honeybee colonies, serve as foraging information centers. Parent birds shuttling back and forth to feed nestlings are followed to food sources by their less successful neighbors. Now we know that seabirds, too, share knowledge of good eating spots, giving and gaining up-to-date information about where feasts may be found, especially when prey is scarce or unpredictable. Ospreys look for other ospreys with fish they've captured from schooling prey that shift within minutes, and fly quickly in that direction, shortening their own search times. Common murres sit on the sea surface a kilometer or so from their nest sites to watch for successful adults carrying fish to their nestlings. This watch zone may act as an information halo, say scientists, where uninformed birds can obtain information on the location of their patchy and mobile prey.

But here's the example that drops the jaw: Guanay cormorants, which live and nest in colonies along the Peruvian coast, form a floating raft of birds that shifts in orientation continuously over the course of the day to indicate the direction of ephemeral prey patches. Every cormorant leaving the colony to fish joins the raft for guidance before heading for feeding grounds.

How remarkable: Birds create an actual living compass for food finding.

Chapter Five

HOT TOOLS

Food is often elusive, ephemeral, patchily distributed, even when it's not scattered over thousands of miles of ocean. As a consequence, birds have evolved not just quirky asymmetrical ears and big olfactory brains, but also remarkably clever strategies for getting at hard-to-get foods, working out tricky problems, using tools—including, perhaps, one of humanity's most quintessential food tools.

Herons all over the world have learned to bait their catches, carefully placing leaves and dead insects on the surface of the water to lure fish. Pied kingfishers, black-crowned night herons, and green herons living in parks have learned that the bread people throw to ducks and geese will also draw minnows. They've taken to plucking up bits of bread and placing them on the water, then waiting with Job-like patience for a minnow to nibble and seizing it with a rapid thrust of their long, sharp bills.

Some birds get at food enclosed in hard shells or other tough packaging by dropping it on pavement to crack it open. Gulls drop clams, and crows and ravens drop nuts. Perhaps most notable is the lammergeier, or bearded vulture, a carrion-eating bird like the turkey vulture, but one that feasts on bones rather than flesh. Small bones it swallows whole; large femurs and ulnas it takes to the sky, letting them go from hundreds of feet over rocky outcrops to split them open and release the marrow. It has its favorite spots for bone breaking, known as ossuaries. This huge and lovely bird is thought to be the one that killed Aeschylus when it dropped a tortoise on the Greek playwright, mistaking his bald head for a stone.

White-winged choughs foraging along the bank of a creek in New South Wales were observed digging up mussels buried in mud inches deep, then opening the more stubborn ones with a tool, an empty mussel shell valve. They held the pointy end down and repeatedly struck the closed mussel until it cracked open. Bristle-thighed curlews in the north-western Hawaiian Islands have been seen picking up sharp bits of coral and slamming them against the large eggs of albatrosses to perforate the thick shells, then sucking out the contents with their long bills. A pair of rangers in the botanic gardens at Canberra saw several varied sittellas probing a hole in a eucalyptus tree with a small twig to remove grubs. Brown-headed nuthatches will do this, too, using bits of bark from long-leaf pines to pry under loose bark scales and devour the edibles beneath. Keen-eyed observers in south Texas spotted green jays using twigs to extract insects from bark crevices in the dry scrub.

A few birds even make their own tools, about as rare a behavior as any in the animal world. Among these is the woodpecker finch of the Galápagos Islands, which selects cactus spines of varying lengths and modifies them to poke into holes and draw out insects beyond reach of its bill, sometimes ferrying a favorite spine from tree to tree to impale its prey. Goffin's cockatoos are wickedly clever at inventing and manipulating tools in the lab, but so far, the birds have never been seen doing so in the wild. (That may change with a new research station established in 2017 on the home ground of the cockatoo, the Tanimbar Islands archipelago in Indonesia.) The Hawaiian crow, or 'Alalā, extinct in the wild, also handles tools in captivity with great dexterity, deftly probing holes with the sticks to pry food out and refining them if they aren't quite right. Scientists suspect it once regularly used tools in the wild.

The most famous tool user of the bird world is the New Caledonian crow, featured in *The Genius of Birds*. This crow makes and keeps its own highly sophisticated tools. It's the only species other than humans to make and use hook tools, little sticks with a hook on the end, which the bird uses to extract grubs and other invertebrates from tree holes and the nooks and crannies of plants. The crows also make very elaborate hooked tools from the leaves of pandanus trees. Pandanus leaves have little barbs along the edge, which the bird uses to latch onto its grubs. It takes a lot of complicated steps to make these tools. The birds make several methodical cuts and tears in the leaf before removing the fully crafted tool from the leaf. This suggests they have an image of the tool in their heads before they actually start making it.

In 2018, the New Caledonian crow showed us that a bird can create tools by combining two or more elements—a feat so far seen only in humans and great apes. For the experiment, researchers caught eight crows in the wild of New Caledonia and brought them to a research station at Oxford University. They presented the crows with a puzzle box the birds had never seen before, which held a small food container behind a door with a narrow open slot along the bottom. They gave the crows long sticks, which the birds promptly inserted through the slot and used to push the food out through a window in the side of the box. Then the scientists gave the crows stick pieces too short to reach the food—some hollow and some solid, with different diameters so that they could fit inside one another.

With no training or guidance, four of the crows put together the pieces within five minutes and used the longer compound pole to reach and extract the food. One bird was able to combine three or four elements to make one long tool—the first evidence of compound-tool construction with more than two elements in any nonhuman animal. This is truly a staggering accomplishment. Children can't make these sorts of multipart tools until at least age five.

The New Caledonian crow's facility with tools continues to defy our assumptions about the limits of avian intelligence. In 2019, researchers showed that these birds could plan several moves ahead while using tools to solve a problem, like a chess player. The experiment found that the crows could hold in mind the type and location of tools that were out of their sight while planning a sequence of tool-using behaviors. This "pre-planning," using mental trial and error, is a key component of human foresight—the ability to form a mental plan before beginning to execute it.

Why does tool use matter? Because it's exceptional in the animal world, and, as ornithologist Alexander Skutch says, "It's precisely this rarity that makes it so instructive. It's evidence of a bird solving a problem or finding a better way to perform a habitual activity."

Some say that when our species learned to use fire as a tool to cook we became truly human; the practice is said to have tripled the brain's processing power. Are birds capable of something as stunning as using fire as a foraging tool?

I once watched from a helicopter hovering above the tallgrass prairies of Oklahoma as land managers from the Nature Conservancy deliberately set fire to vast stretches of grassland to help preserve the largest protected remnant of tallgrass prairie left on earth. The prairie burned fast, fueled by a mat of dead vegetation and fanned by a hard northeast breeze. The air roiled with heat. Prairie grasses—bluestem, switchgrass, and Indian grass—thrive with periodic burning in the spring.

And so, apparently, do hawks that spot the smoky plumes from a distance and home in on the burns. It's something to see. Swainson's and red-tailed hawks, chesty, fierce hunters, wheeling and soaring above the blaze, then swooping low at the fire front to snap up the scores of insects, small ground birds, snakes, mice, voles, and rats flushed out by the smoke and flames. When researchers tallied the number of raptors observed during twenty-five fires in these prairies, they came up with a total of more than five hundred birds of nine different species—seven times the number seen in these areas when no fire is present.

Raptors hunt at fires around the world—in the grasslands and savannas of Australia, Ghana, Brazil, Panama, Honduras, and Papua New Guinea—feasting on the easy pickings of prey fleeing the conflagration.

There's even a word for it: pyric-carnivory. Fire acts as a beater, driving organisms out of the brush. In the savannas of southern Africa, kestrels and jackal buzzards wheel around wildfires and prey on small mammals and reptiles injured, exposed, or killed by flames. Migrating Mississippi kites have been seen feasting on clouds of insects boiling up from summer fires in the grasslands of Texas.

It's one thing to exploit an already raging fire; it's quite another to start one yourself. But at least three species of raptors in northern Australia seem to be doing exactly that. Like raptors elsewhere in the world, fire hawks, as they're called collectively—black kites, brown falcons, and whistling kites—hunt in the vicinity of bushfires. But witnesses have observed these birds doing something radically different: flying into active fires, picking up smoldering sticks, and then dropping them in unburned brush or grass, spreading the flames to new areas, presumably to flush out prey.

Fire-spreading—the transport of burning sticks to spread fire—makes good sense as a foraging strategy. Wildfires draw lots of raptors. Competition is fierce, and there's often not enough fleeing prey for all, so a bird might well benefit from starting a new fire somewhere else and being first on the scene. But if this way of foraging occurs, it's startling behavior that may cross another partition separating humans from other animals. The propagation of fire has long been considered one of those bright dividing lines setting humans apart. "This of course goes deep into our mythos about what separates man from nature, what makes us superior," says Mark Bonta, a geographer and an expert in ethno-ornithology—the study of the relationship between birds and people. "Fire is seen as a sort of divine gift that only humans can wield."

While no unequivocal recorded visual evidence—video or photographs—of fire-spreading in Australian kites and falcons has yet come to light, the behavior is widely known to indigenous people in the Northern Territory, Western Australia, and northern Queensland and confirms long-standing traditional Aboriginal knowledge. When Australian ornithologist and lawyer Bob Gosford moved to the Northern Territory thirty years ago, his curiosity was sparked when he read an account by Waipuldanya Phillip Roberts in a book, *I, the Aboriginal*, published in the 1960s:

> I have seen a hawk pick up a smouldering stick in its claws and drop it in a fresh patch of dry grass half a mile away, then wait with its mates for the mad exodus of scorched and frightened rodents and reptiles. When that area was burned, the process was repeated elsewhere. We call these fires Jarulan.

Waipuldanya's account fascinated Gosford. He had seen birds at fires himself—black kites and whistling kites swooping, diving, gliding in and out of the flames, fluttering about on fire-fueled updrafts. In the Northern Territory, fires are quick and intense, and spread rapidly, especially in the late dry season. Rolling hills of spinifex, buffel, and turpentine grasses, snappy gums, an occasional big eucalyptus tree or eucalypt forest with a dense spear grass understory, create heavy fuel loads. "Very large, long-burning fires can attract thousands of black kites," says Gosford. "Both kite species will go right to the base of the active fire front, within centimeters of very hot fires. Watching these birds foraging at a fire front is riveting—the rush of small birds, insects, lizards, snakes escaping from the base of the fire and the lively pursuit of the birds hunting them."

Beginning in 2010, Gosford teamed up with Bonta to explore whether birds might actually be starting fires in Australia. Bonta was interested in finding explanations that might account for the vast extent of "pyrophilic" landscapes in the world—tropical savannas and other habitats that were formerly burned and are now fire adapted. Lightning strikes alone aren't sufficient to account for them. Geographers and other scholars have long championed the idea that humans were responsible for creating and maintaining fire-adapted landscapes. Is it possible that birds might also have played a role in the evolution of these landscapes?

With the help of firefighter Nathan Ferguson and naturalist Dick

Eussen, Gosford and Bonta began collecting eyewitness reports of birds, especially black kites and brown falcons, picking up burning sticks and starting fires. Plenty of accounts turned up from indigenous people living in the bush, where wildfires are common, and from nonindigenous firefighters and park rangers. Gosford collected twenty reports of the behavior from a dozen ranchers and firefighters with long experience at fire fronts, as well as two academics.

Ferguson, who has wrangled fires for decades in the Northern Territory, said he had observed fire-spreading a dozen times. In a 2018 interview with Gosford and Bonta, he recalled the first time he saw this behavior, in 2001, when he was trying to hold back a fire from entering a radio transmission station on the outskirts of Darwin, the capital city of the Northern Territory. His team of firefighters had conducted a management fire, a "back-burn off," and they thought they had made the area safe, secured the perimeter, gotten the fire under control.

"The kites were really thick," said Ferguson, "and we'd be watching and all of a sudden they'd swoop down and next thing you know we had a fire behind us. . . . The conditions were right," he said. "A hot day, windy, low humidity and dry as a witch's tit, and it just took off like a rocket."

Ferguson saw how the fire got started: One bird dropped a large burning branch over an unburned area, and a new fire flamed up right before their eyes. He stared in disbelief. "I thought, there's no way that a bird swooped down and picked up something that is on fire in a raging grass fire." But after that, he saw fire-spreading behavior multiple times, late in the season, when the extremely dry conditions make for severe fire danger.

A manager of a cattle station and homestead on the western bank of a river in Western Australia described a bushfire that broke out one afternoon on the eastern side of the river. It was a substantial fire, he said, and strong easterly winds were blowing the fire toward the river. He was looking for any embers blowing across the river, and putting out small spot fires. "As the fire burnt to the eastern bank opposite me," he said, "I started to notice that a small number of kites were diving down behind the fire front and emerging with small smoldering sticks, flying over to my side, and dropping them in the buffel grass along the bank." Soon there were numerous small fires going on his side of the river that were beyond his ability to control, and the fire took hold. "Once the fire was raging . . . the kites (hundreds of them) were very active in their pursuit of a feed."

Not all birds that forage at Australian fires know how to start them. When Dick Eussen was working to ensure that a grassfire did not leap a highway, he saw twenty-five whistling kites foraging at the edge of the dying fire. Only two were adept at picking up smoking sticks and dropping them on the unburned side of the road, igniting the grass across the highway. He says he put out seven fires started by the two kites. The behavior appears to occur only under certain conditions, says Bonta, "and it may be only certain individuals or groups that know how to do it and when."

Skeptics dismiss the fire-starting behavior as accidental. Isn't it more likely that the birds are just grabbing sticks on fire by chance as they snap up prey?

Doubtful, say firsthand observers. The kites selectively pick up burning sticks and transport them to unburned locations—the far side of a river or road or an artificial break created by firefighters. "The birds are very aware of what's happening around them," said Ferguson, "and very strategic. They know what they're doing."

Some naysayers point out that there is no known published account of intentional fire-spreading by raptors anywhere else in the world. But perhaps people just haven't noticed the behavior, says Gosford. Or no one has thought to ask the local people whether they've seen it. Or perhaps only hawks in Australia have learned the trick and spread it through social learning.

Studies have shown that birds can pick up novel strategies for foraging within a bird community and pass them around through social networks until they become established behaviors. The classic example of this cultural learning was first noted in the British Isles in the 1920s: A group of great tits, a species known for its superb problem-solving skills, discovered that if they peeled off the foil cap from milk bottles left on doorsteps, they could glean a rich meal from the lovely cream that collects at the top. The tactic began with tits in the town of Swaythling but spread quickly across a wide geographical area, and it wasn't long before tits were stealing cream from doorsteps all across Great Britain.

Ingenious experiments in 2014 and 2015 that aimed to re-create the milk-top scenario confirmed the cultural spread of new behavior in birds. Cognitive ecologist Lucy Aplin trained a few "demonstrator" birds in each of two populations of wild great tits to solve a food box puzzle through two different methods. When Aplin tested the populations in the wild, she found that the vast majority in each of the two populations had learned from the small group of demonstrator birds their specific method for solving the puzzle.

In Argentina, kelp gulls at the Península Valdés discovered a foraging innovation that spread in just this population. In the 1970s, the gulls began feeding on skin and blubber from the backs of resting southern right whales—an awful development for the whales but apparently an important source of food for the gulls. Aerial survey photographs over the next three decades showed that the gull attacks had risen dramatically: Whales with lesions rose from 2 percent to 99 percent, indicating that the behavior had been learned and seeded throughout the gull population.

In learning from other individuals, a bird gets fast, reliable information about foraging strategies without having to invest in trial-and-error learning.

Bonta is aware of old accounts of crested caracara engaging in fire-spreading in Florida, Texas, and Nicaragua. If these accounts are true, perhaps it was a learned skill there, too, that was lost over generations. "As a learned—and possibly taught—behavior, knowledge of it could easily disappear in a bird population," says Bonta. He also points out that examples of tool use by birds in the wild have taken a very long time to verify. To witness and understand behavior that's as rare as fire-spreading would require following a group of pyric-carnivorous birds for a long time in difficult terrain. After all, says Bonta, "it took Jane Goodall and Dian Fossey living with their study subjects for extended periods of time to really revolutionize knowledge about ape societies and to discover previously unknown behaviors." Moreover, he says, "plenty of tool-using behavior that is accepted in the scientific community has not been caught on film. A qualified observer simply noted it and published it."

Some skeptics in the ornithology and fire ecology communities have come around, but many scientists remain unconvinced. "We came up against the idea deeply embedded in the Western mind that Only Man Can Use Fire," says Bonta. "This bias has made it very hard for people to take seriously forty-thousand-plus years of accumulated knowledge of the landscape accrued by Australian Aboriginal people. Every group in the Northern Territory and probably elsewhere in northern Australia insist that fire-spreading is a real thing."

Gosford and Bonta are now exploring the correlation between fire-spreading behavior and environmental conditions. They suspect that fire-spreading occurs more often in the rainy season, when fires are scarce. This indicates that the birds likely play no role in the increasing frequency and severity of bushfires in Australia. "If grass is thick and no fires are occurring or are small in area, birds may seek to start new fires or spread existing fires themselves," says Bonta. "This even occurs with cooking fires—we possess several reports of the well-known phenomenon of birds grabbing embers and dropping them on unburned fields nearby."

If it exists, fire-spreading may be evidence of highly sophisticated cognition in these raptors—a grasp of a two-step sequence of cause and effect: first, that dropping smoldering sticks will set a field afire, and second, that the resulting fire started in a new spot will drive out prey. Moreover, if kites and falcons have learned to manipulate fire to flush out prey, this would be the first known use of fire as a tool by any nonhuman animal, overturning some old entrenched assumptions—the orthodoxy that only humans spread fire, and that our singular mastery of fire was largely what made us masters over the environment. But it could have worked the other way around, says Bonta. "It could very well have been the case that humans, birds, and fire coevolved in some sort of mutualistic relationship—perhaps humans actually derived the idea of using fire from watching birds. This is what plenty of myths about fire-spreading among indigenous peoples in Australia and elsewhere in the world tell us, after all—birds were the original fire-starters."

I love this idea: that a bird with a burning stick might upend old Promethean notions of human uniqueness and ecological mastery.

Chapter Six

TRACING THE
ANT'S PATH

A world away, in the rainforests of Costa Rica, some clever neotropical birds have found another remarkable tool for finding prey. It's perhaps not as flashy as fire, but similar in effect and requiring equal ingenuity, along with a firm grasp of the behavior not of an element, but of another kind of animal.

You often hear it first. A sound like the pitter-patter of rain beginning, a crackling and snapping, especially if the leaf litter is dry. And sometimes, the high whine and buzz of flies, and a noisy, boisterous chirring, chittering, grunting, snarling, hissing, whining, and squawking.

Then you see the cause of all the fuss. A seething fan-shaped mass of tens of thousands of army ants boiling across the forest floor. As the raid nears, the noise increases. "You hear the sounds of the critters fleeing from them—harrowed roaches, katydids, crickets, scorpions, big-bodied insects flying, running, hopping to get out," says biologist Sean O'Donnell. "They're like animals fleeing a wildfire."

The ants are *Eciton burchellii*, the mini-lions of Costa Rica's neotropical forest, ferocious pincer-jawed predators that capture nearly every arthropod in their path. They're pouring out of a huge temporary nest site known as a bivouac. The nest is made of the bodies of live ants themselves linked to one another by their own limbs, creating a huge, jiggling, thermoregulated sanctuary for the queen and her larvae.

The raid spreads out from a bivouac over the jungle floor, covering every nook and cranny, savage and relentless, scaling trees to attack wasps' nests, subduing and dismembering large-bodied insects, biting and stinging snakes and other vertebrates, overwhelming even the most vicious scorpions. The prey the army ants take on these raids, as many as 30,000 corpses a day, are carried back in braided lines of workers to feed the 60,000 to 120,000 larvae in their bivouacs.

"The ant colony sends out only a fraction of its worker force because it needs to maintain the integrity of the bivouac," says O'Donnell. "But the raids can be spectacular, a carpet of more or less continuous ants fifteen to thirty feet wide, advancing through the forest at something like fifteen yards an hour."

The chirring, chittering, growling, and caroling are the loud vocalizations of neotropical birds, which follow in the raid's path. As the ants progress, the moving wave flushes swarms of insects that try to fly, hop, leap into the nearby undergrowth or crawl up the trunks of saplings, dishing up a fluttering, panicky banquet for the birds.

A huge diversity of birds attend these ant raids to find their daily food—as many as a hundred species at a given location. Among them are species of rainforest birds so dependent on the ants and so specialized to find and follow them that the tiny insects are built right into their names: ocellated antbird, bicolored antbird, spotted antbird. These, along with certain woodcreepers—the plain-brown and the northern barred—get most or all of their food from ant raids. But more opportunistic birds also frequent the raids, among them various other resident species of antbirds, including the chestnut-backed, as well as tinamous, tanagers, and motmots and occasional migrants such as thrushes, warblers, and vireos familiar from temperate forests—the Kentucky warbler, Canada warbler, and Swainson's thrush.

None of these birds are silly enough to eat the ants themselves. Rather, they mooch off the ants' labors, swooping down and picking off creatures the ants have rustled out of hiding. It's a good example of kleptoparasitism, that form of freebooted feeding in which one animal takes food that was caught, collected, or prepared by another.

Other kleptoparasitic birds horn in on the prey captured or stirred up by other species—those drongos that use mimicry to snatch prey from meerkats and babblers, for instance, or double-toothed kites that shadow squirrel monkeys in these same Costa Rican forests, snapping up the insects and small vertebrates flushed by the monkeys. The azure kingfisher of Australia and New Guinea—a gorgeous bird with a royal-blue head sheened with violet—perches above foraging platypuses and dives for fish, crustaceans, and frogs disturbed by the mammal, then swoops back up to its perch to slam its prey against the wood before swallowing it headfirst.

But for sheer numbers of pirated prey, ant-following birds are hard to beat. Each day at a single raid, birds steal around two hundred large prey items from the tiny ants—around a third of the ants' daily take—eagerly snatching up katydids, roaches, spiders, and scorpions. The ants resist the birds' efforts, attempting to protect their food with large soldier caste workers known as majors, which drag large prey items out of sight, hiding them under the leaf litter or stashing them away in food caches along the foraging trail.

Birds rarely steal booty directly from the clutches of the army ants—but not because they are at risk of becoming prey themselves. These ants are not like their counterparts in Africa, which have hooked mandibles with a knifelike edge designed for cutting through skin and flesh. According to O'Donnell, African driver ants can attack, kill, and eat vertebrates as large as cattle, antelope, and even humans—if they're drunk or asleep. And babies. "They have enough workers to pull that off," he says. E. O. Wilson tells the story of allowing driver ants to sweep through his critter-infested house in Gorongosa National Park in Mozambique on a regular basis. "You just go away and have a cold drink somewhere,"

Wilson says. And after a few hours, the army has passed through, massacring and taking with it every animal it finds. Then "you can return home—a home that has been perfectly cleaned for you."

While the new-world army ants of Costa Rica don't have the cutting mandibles of African driver ants that can flay flesh, they're still fierce predators and will do the sort of "housecleaning" Wilson describes, sweeping through rural houses in the lowlands near forests, clearing out any living arthropod—roaches, spiders, wasps. They also sting, and their toxic venom can be fatal to birds. Army ant venom contains proteolytic enzymes that start to digest their arthropod prey and break it apart, dismember it, so the ants can take the parts back to the colony. "For us, an army ant sting is just painful and unpleasant," says O'Donnell, "but to a bird that weighs a few grams, it can be lethal." Ecologist Johel Chaves-Campos, who studies ocellated antbirds, once saw a juvenile dying after getting stung in the face by three worker ants when it was tangled in a mist net.

Why would birds bother with such a risky foraging strategy? You would think insects would be a dime a dozen in the rainforest. But the payoff is worth it. The big juicy katydids, roaches, and other hefty insects preferred by tropical forest birds are tough to catch—mobile and typically active mostly at night, when birds are sleeping. The ants concentrate the feast. An ocellated antbird can glean about fifty tasty insects or arthropods in a few hours of feeding at a raid. It would take the bird more than twice that long out there alone in the tropical forest trying to rustle up its own buffet.

Not that ant following is an easy feat. On the contrary. Only lately have scientists discovered just how difficult and complex an effort it is, requiring sophisticated mental skills—learning, remembering, sharing information, and perhaps even planning for the future.

The relationship between tropical birds and the ants they stalk has been known for more than half a century. But just in the past decade or so have researchers studying ant-following birds uncovered some of the subtleties and nuances in this strange form of foraging. Their parsing of the bird-ant liaison is challenging old ideas about simple competition at the swarms and turning up new ones about how birds "read" the ants and what sort of intelligence is involved. Antbirds, it turns out, may act less like gulls squabbling over a sandwich and more like a shrewd wolf pack tracking elk. Their memories may have more in common with the intricate remembering of an elephant than a reptile. And like honeybees and cliff swallows, they may act as foraging information centers for a host of other bird species.

"These raids are so inherently exciting, complex, *beautiful*," exclaims O'Donnell. That word *beautiful* most often refers to things that we can see—visions of rainbows and resplendent feathers. But O'Donnell is not talking about that kind of beauty. He's talking about an invisible allure apparent only in the hidden exchanges between creatures and the secret understanding a bird may have for the patterns of the insects it follows.

A biologist at Drexel University, O'Donnell spent eleven years in Costa Rica, from 2005 to 2016, observing more than seventy army ant raids and the birds that attend them in an effort to understand their interactions. "Every raid is different," he says. "In my time observing, I've gotten a sense for the raid patterns. But I could go down to Costa Rica right now, spend a couple weeks watching the swarms, and I'd see something I've never seen before, something beautiful and elegant. There's a lot going on and a lot that we still don't understand."

One raid in particular is deeply embedded in O'Donnell's memory. He is a member of the "white fang club," as one reporter put it, an exclusive group of biologists who were bitten by a highly venomous snake and lived to tell the tale.

It happened more than a decade ago, when O'Donnell was with a group of students tracking a swarm of army ants in the lowland rainforest of La Selva Research Station, a small reserve in northeastern Costa Rica that O'Donnell says is renowned for being a "predator-infested pit," in the words of a fellow researcher. The ants were bivouacking in a tree right next to one of the main trails. In the lowlands, ants often form a bivouac in a tree cavity, sometimes high up on the trunk. This one was about thirty feet off the ground. The ants were pouring out of it, down the tree, and off into the forest.

"The game with army ants is that you follow this narrow, linear column until you come to the raid front, where it fans out into this big mass, and that's where the birds are," says O'Donnell. He and his students were tracking the column through the forest looking for the raid front, when they came to a difficult spot. "Army ants like places that humans don't like," explains O'Donnell. "They go into the nastiest, most difficult sites, like treefall gaps, which are tangled and messy and often full of snakes."

O'Donnell didn't want his students traipsing into the gap, so he went into it himself to see if the ants were coming out the other side. And sure enough, he found the line fanning into a raid front attended by a flock of antbirds.

"I lifted my binos to look," he recalls. "And then, *BAM!* Oh my god! It was like a sledgehammer with a nail on it had gone hard into the heel of my left foot." O'Donnell ripped off his thick rubber boot and sock and saw a single bloody wound—but he didn't see the scorpion or bullet ant he thought might have stung him. "The pain was insane. I've been stung by a lot of things, and some of them hurt, but this was off the charts."

Then things got worse. O'Donnell started hallucinating, stumbling, going in and out of consciousness. "I wanted to grab a sapling to hold me up, and I tried to focus on this one little sapling in front of me, but it looked like there were five or six of them. I wasn't seeing double; I was seeing sextuple." He made it back to the station, but the bite was misdiagnosed as a scorpion or spider sting and treated with a little salve. After more than twelve hours, however, O'Donnell was peeing thick black blood and was pretty convinced he was going to die. Finally, he was taken to a hospital. When the doctors came back with the results of his blood test, they said nothing, just grabbed him, ran him down the hall to a bed, jammed an IV in his arm, and started a drip of antivenin. "They later told me that the time to coagulation in my blood was, well, infinite," he says—in other words, he had *no* coagulation factor, and the doctors feared brain bleeding and damage. Ten vials of antivenin later, he was vastly improved. But exactly what had struck him remained a mystery until he got home, and his wife asked if he still had the boot. When she picked it up and crinkled it, another fang hole opened up, the telltale two-fang mark of a snake. "One fang had penetrated but the other hadn't," he says. "That's what saved my life." The snake was probably a bushmaster, which means that O'Donnell was very lucky indeed. The fatality rate for a bushmaster strike is about 50 percent.

"That snake was probably hiding in the leaf litter, minding its own business," says O'Donnell. "Bushmasters are nocturnal hunters and stay quiet in the day. But then this army ant raid comes through, covers it, stings it, bites it. The poor snake was probably annoyed by the ants, in a pissed-off mood. Then I came stumbling along." For years, O'Donnell couldn't go back into the area of that treefall gap at La Selva. To him, it was as if the snake, or at least its ghost, was still present, so visceral was his response to the place. "I'd walk along that trail and I would just feel this evil presence. Completely superstitious, but very real."

The experience doesn't seem to have dampened O'Donnell's enthusiasm for ant swarms and the birds that attend them. "The raids are incredibly exciting," he says, "and far more subtle and complex than we imagined." For one thing, they may not be the hotbed of competition and exclusion they were once thought to be. The traditional view is that raids are antagonistic affairs, with birds ferociously jostling for resources. The ocellated and other obligate antbirds like to keep the ant colonies to themselves, it was supposed, fending off any other birds that might horn in on their finds. According to Edwin O. Willis, who wrote about antbirds in the 1960s and 1970s, ocellated antbirds plant themselves in the best foraging spots and drive out the smaller birds with "loud snapping of the bill and a whirr of wings on an arrowlike attack. As a consequence, many competitors move away as soon as the antbird approaches."

O'Donnell doesn't see it this way. He argues that birds attending ant raids are not so much competing as coexisting, sharing resources, perhaps even facilitating one another's participation—deliberately or not. "Having other birds around may be an advantage because there are more eyes and ears to detect predators," he says. "Food is usually abundant at a raid, so the costs of allowing other birds to attend may be low, favoring more positive interactions."

Anyone who has ever watched a duel between two rufous humming-birds at a nectar feeder, or a pack of gulls battling for a stray french fry, or a pair of starlings bloodying each other over just about anything knows the potential for fierce competition in the bird world. The common un-derstanding of the struggle for survival is that it's a fight to the end to win limited resources. As the Dickens character Noddy Boffin famously says, one must either "scrunch or be scrunched."

But there are plenty of examples of cooperative foraging in the bird world. Pelicans synchronize their movements, forming a line or semicircle of a dozen or more birds around schooling fish, driving the fish to concen-trate in shallower water so they're easily scooped up from the sides. Gulls, terns, boobies, and cormorants do something similar in mixed flocks. As the size of the group builds, so does their hunting success. Raven pairs hunt in tandem, working a seabird colony the way pickpockets work a beach crowd: One raven flies at an incubating adult to divert it, while the other swoops in to seize the exposed egg or chick. Harris's hawks in New Mexico hunt together in family packs of up to six members, like winged wolves, to take down prey too large for them to dispatch alone, most often a cottontail or jackrabbit. This kind of coordinated hunting by multiple pack members to capture and share large prey has been seen only in a few mammalian carnivores—gray wolves, chimps, dolphins, lions.

"When I watch the interactions at ant raids, I don't see birds repuls-ing other birds," says O'Donnell. There's more than enough food for ev-eryone. "The raid is so physically large and there's so much prey that no one bird, or even family of birds, can really monopolize it. All the birds can stuff themselves to their heart's content."

Moreover, the different species of ant-following birds partition the foraging space at swarms into horizontal ground zones and vertical strata,

forming "guilds," or groupings that exploit their own niches using different strata and different strategies. At lowland raids, black-crowned ant-
pittas and rufous-vented ground-cuckoos situate themselves away from
the swarm front, feeding on the ground at a distance. Birds like the plain-
brown woodcreeper and gray-headed tanager use tree trunks well above
the swarm. The "perchers" or clingers such as ocellated antbirds, bicolored antbirds, and spotted antbirds perch in the vegetation directly over
the raid. "They peer down at the swarm," says O'Donnell, "and when
they see something valuable, they'll pounce and then quickly fly back up
to consume their prey. A precision strike, as it were."

The most dominant of the perchers in Central America is the ocellated
antbird, studied in detail by Johel Chaves-Campos during his observations of more than two hundred ant raids. Beautiful, charismatic, with a
striking scalelike feather pattern tinged in gold, and bright blue facial
skin, the ocellated antbird often occupies the choice spot for foraging, a
perch low to the ground at the leading edge of the swarm. The consummate professionals, ocellateds are designed for the ant-following business.
With strong, thick legs and curved claws, they are adept at gripping thin,
vertical branches and stems and swinging to position themselves perfectly
for a quick sally to the ground. They hop and flutter between perches,
looking for anything juicy of reasonable size flushed by the army ant
swarm, including small lizards and cockroaches up to two inches long.

Then there are birds that actually walk on the ground among the ants.
"Which is kind of amazing," says O'Donnell. There's real danger to the
birds if they get stung by army ants. "These birds don't hang out on
the periphery and dart away," he says. "No, they're right down there in
the center of the swarm. But they're extremely adept at dodging the ants
and mostly avoid getting stung." Moreover, as Chaves-Campos points
out, it's not easy for an ant to inject venom into the flesh of a bird's foot
because it's covered with scales. The ocellated antbird that was stung and
died in the mist net got injected with venom because the ants were able to
reach the bird's face, which has no feathers, he says. "Under natural conditions, the birds can remove ants from their feathers or feet before they

can reach an area of bare skin." An antbird reacting to the occasional bite or sting stamps or shakes its foot, pecking at the offenders and tossing them into the air, "jiggling or jittering back and forth from one foot to the other, dancing about as if on hot coals," as Edwin Willis described it.

A rmy ant raids may offer birds a rich food bonanza, but finding the swarms is tricky. The colonies are hard to locate, erratically distributed throughout the forest, and well concealed in the dense growth of the shaded understory. Encountering a swarm by chance is unlikely, and it's easy to miss one just a few hundred feet away.

Furthermore, the raids are sporadic. Antbirds can't count on predictable daily swarms. As the army ants raise their young, colonies alternate between periods of high and low raiding activity. It all depends on the development phase of the brood. During the two-week period when the ant larvae are growing, the babies have a voracious appetite and must be continuously fed, so the ants swarm intensively, starting their raids just after sunrise and going strong for a full seven hours. This is their nomadic high-raiding phase. Each night, they move their bivouacs to a new location, on tree roots or buttresses, so they can cover new ground with a fresh batch of prey each morning. Then, when the larvae pupate (a nonfeeding stage), and the queen lays more eggs (which also don't require food), the colony enters a stationary, low-raiding phase, and the ants remain in the same bivouac site, usually inside a hollow tree, for three weeks, swarming only occasionally in a random pattern and for a shorter period of time.

This on-again, off-again cycle makes it challenging for birds to track the ant raid activity and movements over days. But antbirds have been following ants for a very long time—five million to six million years—and they have learned the ants' patterns. They rely on the visibility of the bivouac on those tree roots or buttresses during the nomadic phase and its predictability during the concealed stationary phase. Also, more than twenty species of antbirds, including the ocellated and the bicolored, have evolved an ability to track the colonies in space and time and anticipate the swarms across their cycle. The skill was first described by biologist Monica Swartz from observations she collected in the Costa Rican lowland forest. It's called bivouac checking, and it's a truly remarkable feat.

To find out just what these expert antbirds are doing, Johel Chaves-Campos followed individual radio-tagged birds at La Selva, mapping their position every ten minutes or so from the moment they left their roost. He held "bivouac vigils" sunrise to sunset, hiding in an inconspicuous spot around twenty feet from the ant nests. After more than seven hundred hours of tracking, he found that the antbirds visit multiple ant colonies in a single day to keep track of different bivouac and raid locations and what the ants are up to there.

Twilight in the jungle, still light, but evening is descending. Fully satiated from a day of feasting, an ocellated antbird follows the trail of army ants back to their bivouac. It looks around, inspecting the location and encoding it in memory. If the ants are in the midst of moving to a new bivouac site, the bird follows the immigration column from the old site to the new site. "In a sense, it's putting the ants to bed before returning to its roost," says O'Donnell, "which is what researchers do."

Early the next morning, the bird flies directly to the bivouac, lands on a nearby perch, sometimes only inches from the ants, and peers into the bivouac and the ground around it to assess the ants' activity. If the ants are raiding, the bird will track the raid to its front and forage there; if there is no activity, it will move on to another bivouac to check its raiding status.

It sounds straightforward enough. But as researchers have learned, keeping tabs on a bivouac is not easy. "The ants do sneaky things," says O'Donnell. "You think they're going to bivouac in a certain spot and then you go back the next morning and they're not there. They emigrate to a new site, and then they emigrate again, sometimes after dark." If the bivouac was moved at night, and a bird can't find it, it will search along the path of the previous day's raid for the ants' new nest or raid location.

Sometimes, the searching birds will go back to an abandoned bivouac site again and again, for several days in a row. "It's like there's a memory trace they can't erase," says O'Donnell. "I've even seen hard-core followers, obligate antbirds, repeatedly going back to bivouac sites no longer occupied. They do all the things they would do if the ants were there: They get very close, they orient to it, they peer inside. They perch there and get kind of forlorn looking. Sometimes you watch animals and you can't help but feel like they're expressing human emotion. 'Disappointment!'"

Revisiting bivouac sites and exploring them thoroughly is actually very reasonable. It's worth keeping track of these inactive sites in case the ants have not moved at all but instead entered a stationary phase, says Chaves-Campos. Moreover, it's worth looking closely because the ants are good at hiding. Sometimes they'll bivouac under a rock or a hollow log or in the buttresses of a live tree. At high elevations, they'll even go underground in a little tiny hole where you can't see them at all. "We've been fooled by this sometimes," says O'Donnell. "We'll look and think, 'Oh they're gone.' And then when we look more closely, we'll see a couple of workers and realize that the bivouac has gone farther into the cavity and is just no longer visible. Assuming that birds can make the same mistake, it makes sense that they check very carefully because it's not immediately obvious that the ants have actually left." So there may be an adaptive explanation for going back to former bivouac sites; occasionally it pays off.

But O'Donnell also wonders if it's similar to how he thinks about the last place he had chocolate. "Maybe there's something deeply satisfying about being at a place where the ants have been, a kind of echo of reward. It's not uncommon in cognitive systems." Including ours.

Not long ago, on a long drive along a country road, I stopped at a convenience store to stem my hunger. On impulse, I bought a bag of peanut M&M's. I had not had the candy since college, more than four decades earlier. As I drove along, popping the pieces in my mouth, I was transported back to the tiny cubicle on the fourth floor of my college library, where I used to fuel all-nighters with a big bag of the candies. I could almost smell the musty books, see the bright little desk lamp, feel the hard wood of the cubicle chair and the absolute necessity of those choco-late peanuts to fight the heavy fatigue.

Food triggers memories not just of the stuff itself but where and when it was eaten. There's a good reason for that. So important to survival are deeply rewarding high-value food events that the brain prioritizes their recall and stores them in a privileged place. The process may involve do-pamine, a chemical known to be important for rewards, and the hippo-campus, the brain region critical to long-term memory in both humans and birds.

Through painstaking research, Johel Chaves-Campos confirmed that ocellated antbirds were capable of holding in mind the locations of multiple colonies, as well as their raiding status. At a bare minimum, the ocellated antbirds are memorizing where the ants are. They're probably also checking whether a raid has started and, if it has, what direction it's heading in. And they may be doing much more than that.

Chaves-Campos noticed something else: An ocellated antbird tended to move in a straight line between bivouac sites in the rainforest, like it knew precisely where it was going, even if it hadn't actually visited the destination itself. Just before it left a bivouac site, the bird would sing a loud song, probably directed at its mate, which acts as a "departure sig-nal," says Chaves-Campos. The other ocellated antbirds present at the site, eavesdropping on the signal, would depart, too, and travel silently through the rainforest, moving single-file directly to a new site, each bird separated by a few seconds like orderly kindergarteners on a city walk. No meandering in different directions, just a beeline from one site to the next, as if one or more of the birds in the group already knew the exact location.

Which, it turns out, they do. In one instance, an antbird in a group traveled directly to a new nomadic ant colony that this particular bird had not itself inspected on the previous evening. Other group members *had* scoped it out while the ants were moving their nest overnight to a new location, one hundred meters away from the old one. In each of the three trips from site to site, all of the birds in the group traveled directly and arrived simultaneously, within ten seconds, at the new site.

"This suggests that these birds do not wander looking for ant colonies," says Chaves-Campos. "They know where to go, or at least one individual in each travel group knows the location of the colony. By following those who know, the antbirds discover ant colonies previously unknown to them."

In other words, bivouac-checking birds may be sharing information, acting as heralds of rich food sources for other birds. And not just for fellow specialists, but for other foraging birds, as well. In an experiment with recorded "playback" vocalizations, Chaves-Campos found that the calls of specialists like the ocellated and bicolored antbirds attracted generalists, too, suggesting that these other antbirds may use the calls of the experts to help them locate swarms without having to search themselves. Like common murres and cliff swallows, antbird flocks may be like little roaming centers of food information and intelligence.

This dovetails with O'Donnell's notion of a foraging system that's not necessarily as cutthroat competitive as once imagined. "The birds may actually be facilitating each other's attendance at the raids," says O'Donnell. "When colonies move to a new location, that first day the specialized set of antbirds tend to be at the raid alone. But then the next day, the flock is completely mixed, generalists and specialists together, and the specialists are not fighting off the other species. Now, is this cooperative? Is it intentional? We don't know. But I would argue that at least it's commensal."

What happens when these professional ocellated antbirds aren't around to run the show? The island of Barro Colorado in Panama provided a natural experiment for Janeene Touchton of Princeton University and her colleagues. The island had lost its population of ocellated antbirds when the birds vanished following a harsh dry season in 1969. Over the following decades, the smaller, more subordinate spotted antbird assumed the mantle of the bigger, now absent ocellated antbirds, more than doubling its population density and shifting its use of the swarms, moving from operating as a somewhat more opportunistic ant follower, foraging at swarms only when they passed through its territory, to spending most of its foraging time in and around the swarms and interacting with other birds the way an ocellated antbird does—a surprising show of behavioral flexibility.

O'Donnell believes that ant-following birds may be capable of something perhaps even more impressive than information sharing or behavioral flexibility: mental time travel. This is the ability to travel in the mind back into the past to remember the specifics of past events—the what, when, and where—and to use this memory to plan for future actions. If it's true, ant followers would join an elite group of birds (along with a few primates) that have lately challenged the notion that mental time travel is unique to humans.

It's worth pausing to consider the nature of the skill.

Looking at some relatively recent human endeavors—introducing kudzu to control erosion or cane toads to control beetles or the endless dithering over what to do about climate change, it would seem that our species can be fairly inept at using the past to plan for the future. But most of us have at least some capacity for it. We're not born with it. The ability to remember the specifics of a past event and to envisage future episodes emerges in children between the ages of three and five.

Psychologists Michael Corballis and Thomas Suddendorf suggest that the human capacity for mentally wandering backward and forward in time may have evolved in our ancestors around 2.5 million years ago, when the global shift to a cooler, drier climate transformed the woodlands of southern and eastern Africa into open grasslands. "This left the hominids not only more exposed to attack from dangerous predators, such as sabre-tooth cats, lions, and hyenas, but also obliged to compete with them as carnivores," they write. "The solution was not to try to compete on the same terms," but rather to carve out a new "cognitive niche" that relied on the accurate recording of specific events in the past (encoding information on "who did what to whom, when, where, and why")—and using this learned information in planning for the future. This capacity for mental time travel, the theory goes—more even than our opposable thumbs or talent for language—is what gave wings to the human mind and imagination and is now considered a defining quality of *Homo sapiens*.

Until quite recently, scientists thought ours was the only species that could do this sort of backward-forward thinking. Other animals existed in the tyranny of the present, a kind of eternal now. They didn't travel back in time in their own minds because they didn't need to. That view changed with Nicola Clayton's inspired experiments on western scrub jays in the 1990s. Clayton and her colleagues showed that scrub jays have a powerful ability to remember the past and use this information to plan for the future. Scrub jays manage to recall not only where they buried their food, but also what they buried in a particular spot and when they buried it there so that they can retrieve first those foods that are quick to spoil, like fresh fruit, insects and worms, and save for later those that are more enduring, like nuts and seeds.

"It's a poor sort of memory that only works backwards," says the White Queen to Alice. It turns out that these jays have the sort of memory that also allows them to travel forward in time.

Scrub jays steal from one another's caches. Clayton and her colleagues found that jays that know they are being watched by another bird will later move their caches, presumably to prevent them from being stolen. But they will do this only if they themselves have pilfered the caches of other birds in the past. So the jays are drawing on their memories of their own past thieving exploits and linking them to the possibility of future theft by another bird, then using this time travel to modify their caching behavior and defend their stores from pilferers.

Other caching birds are memory artists, too. They have to be. If you're a bird that stores your food in caches for future use and you want to retrieve what you've hidden, what's your best strategy? You might try looking in all the likely hiding spots—the same way we might look for spare change in the pockets of our jackets or coin dishes around the house. But this sort of random hunt is an inefficient way to satisfy real food needs—just as poking around in old pockets is an ineffective way of coming up with significant cash. Far better to recall where you actually put your stashes. Caching birds such as pinyon jays, black-capped chickadees, and Clark's nutcrackers are capable of remembering the spatial locations of thousands of their own hidden food stores and recovering them with uncanny precision even months after caching—and even though the landscape may have changed with shifting soil, rock, and snow. (In fact, when the nutcracker is burying its seeds, it smooths the ground to remove signs of disturbance.)

Scientists suspect that caching birds remember where they hid things by creating specific visual memories of the relationship between the cache and a range of visual and spatial cues, especially conspicuous landmarks, some nearby, like trees, stumps, and rocks, and some more distant, like mountains and mountain ranges. Still, it boggles the mind to think that a bird with a nut-size brain could hold in its memory so many specific hiding spots over such long periods of time.

Equally impressive are the keen spatial memories—coupled with sharp timekeeping skills—of foraging hummingbirds. If you're a hummingbird, keeping up with the energy demands of your little thrumming, hovering body is no mean feat. It requires visiting hundreds of flowers a day to feed on their nectar. You can't waste energy revisiting flowers you've already sucked dry. When a hummingbird drains nectar from a flower, it takes time for that flower to replenish its supply. Visit the same bloom too soon, and it will still be empty; wait too long, and a competitor may make off with your precious drink. You must remember what flowers you've visited and when—in fields with thousands of flowers. It's like a giant game of Concentration.

As it turns out, hummingbirds are champions at this. Field studies show that the tiny birds can remember not only the spatial location of a particular flower but also when they visited that flower, along with its nectar quality and content (different flowers vary in their sucrose concentration), and how quickly its nectar refills. At the University of Edinburgh, Sue Healy and a team of researchers created fake flowers stoked with sugar. One type of flower they refilled every ten minutes and another type, every twenty minutes. The hummingbirds quickly learned how long to wait before coming back to each type.

The birds are not pinpointing flowers in a meadow in the way you might think, using the color or look of a flower as beacon cues. Healy's team shuffled the location of their fake flowers and discovered that the birds pin their memories strictly to location. They watched a male hummingbird fly toward the location of a flower he has previously visited. As he approaches the location, he stops. The flower is no longer there. "He moves closer, hovering in three-dimensional space, rotating his whole body to scan the scene," write the researchers. "Still no flower. After a few seconds, he departs to look for sustenance elsewhere, apparently failing to notice that the flower from which he was expecting to feed is still in the meadow. The flower looks the same. But it has been moved 1 meter from its previous position."

Just how hummingbirds recall the sites of particular flowers is still largely a mystery. But the researchers suspect the birds may relocate a flower through so-called view-matching, scanning into their memory views of a place from a flower—either a kind of snapshot of a landmark or two, or a more panoramic view of patterns of light, color, and motion—and then using these views later as a reference, matching their current view to the remembered visual "snapshots."

Skill at these memory gymnastics is absolutely vital for sustenance, but it may also influence mating success. The ability to remember the what, where, and when of nectar availability in daily foraging, it turns out, is key to the dominance and mating success of at least one kind of hummingbird—even more so than its body size or weaponry.

A common neighbor of the antbirds in the Costa Rican rainforest, the long-billed hermit hummingbird is big as far as hummingbirds go, about twice the size of a ruby-throated, with a long, curved bill for sipping nectar from heliconia flowers. Like the devious little hermit hummingbirds of Trinidad, male long-billed hummingbirds form leks in the forest understory, where they sing and display for up to eight hours a day. They do this for the entire eight-month breeding season, an extremely demanding effort sustained only by males in excellent physical condition. Dominant males battle over singing perches, sometimes attacking each other with their sharp, pointy bills.

Females visit the leks just once per day, so males must be present and prepared to mate throughout the long breeding season. If they leave to tank up, they risk losing their coveted singing perch, so it pays to know where the nectar-rich flowers are among the thousands of forest blossoms—in order to get there and back quickly. The cost of a single error is probably minimal, says Marcelo Araya-Salas, who studies the hummingbird, but the cumulative cost of making multiple errors during multiple foraging trips each day over 240 days could set a bird way back in its intake of calories.

To explore whether there was a connection between the hummingbird's spatial memory and its ability to acquire and defend a lek territory—a critical element for mating success—Araya-Salas and his colleagues tested how well males could remember the locations of rewarding hummingbird feeders he placed near the leks. Birds that scored best on these tests turned out to be the dominant males with prime perches at the lek. When it came to holding and defending mating territory at the lek, a good memory for food locations beat out all the advantages of physical prowess—bigger body, bill tip size, even flying power.

After some extensive observations of bivouac-checking antbirds in Monteverde, Costa Rica, with his colleague Corina Logan, Sean O'Donnell suspects that bivouac-checking antbirds face challenges a lot like those of food-caching birds and hummingbirds—and are capable of similar mental feats. The antbirds have to remember the location of the bivouac so they can return to it the next morning (the what and where of episodic memory), while also remembering which nests are in the nomadic phase (the when). They must also get there before the ants start their raid (more when). Clearly, an antbird will do better if it can keep track of the locations of multiple colonies and if it can remember whether particular colonies are in nomadic, high-raiding phases.

Moreover, bivouac-checking birds show signs of behaviors that may signal planning. When an antbird checks a bivouac in the evening after a day of gorging, it does not feed; it's just gathering information. Only in the morning, when the bird returns to the location it discovered the night before and finds the ants swarming again does it encounter its reward—the bounty from a fresh raid. In other words, the antbirds are not responding to their current state (satiation), but rather a future need (breakfast!), which suggests they may be anticipating a future event: "Check location of bivouac now in anticipation of foraging the next morning."

The memory skills of caching jays and bivouac-checking antbirds may pale compared with the sophisticated human ability to mentally wander forward and backward in time. But who knows? Maybe that especially juicy cricket picked up at an outstanding ant swarm is an antbird's version of the proverbial Proustian madeleine dipped in tea. Maybe there are places, an abandoned bivouac site tucked into a tree root, or a patch of forest exposed to hawks and falcons, that carry for an antbird the same powerful memory that snake-infested treefall gap or his favorite spots for chocolate carry for Sean O'Donnell—a haunting echo of danger or sweet reward.

Or perhaps, only perhaps, says Chaves-Campos, scent plays a role. When he worked at La Selva, Chaves-Campos noticed that the abandoned bivouac sites smelled like *Eciton burchellii* for several days and was even able to locate some of the ant swarms by tracing the odor. Perhaps, like petrels homing in on the odors of their ocean prey, antbirds have evolved the ability to cue in on the scent of the streaming insects so vital to their sustenance.

Play

Chapter Seven

BIRDS OF PLAY

Mathias Osvath remembers watching a pair of ravens soaring above the rolling wheat fields of southern Sweden, rising, falling, swooping. Suddenly, one bird folded its wings to its side and plummeted straight down. "It dropped like a shot duck," he says, "And then boom! Just before hitting the ground, it shot out its wings and recovered." Another time, he saw a raven walking along the ground in that strutting way ravens have, and then out of the blue, it toppled over on its side. "The first time I saw this I thought, 'Falling over like that for no apparent reason? That bird must have some kind of neurological disease.'"

But this wasn't palsy or paralysis; it was play—Osvath's first glimpses of it—by one of the avian masters of the form. The odd, goofy behavior with no apparent purpose in a bird renowned for its seriousness and intelligence struck Osvath, a cognitive zoologist at Lund University, as remarkable and worth investigating.

For most of us, *playful* is not what pops to mind for the bird Edgar Allan Poe called that "grim, ungainly, ghastly, gaunt, and ominous bird of yore." With their black cloaks and croaking, ravens are more closely linked with foreboding and death. The name for a group of these birds? An "unkindness" of ravens—and for good reason. Raven society is rife with conflict and fighting. When the birds are young, they can live together harmoniously, but once they pair off, they become extremely territorial and aggressive and can actually kill one another. Renowned as ferocious hunters, ravens are the "black pirates" that rob the nests of other birds and peck the eyes from newborn lambs, "indiscriminately voracious," as ornithologist Edward Howe Forbush describes them, quick to take advantage of "anything edible, alive or dead, which it can catch, kill, disable or pick up: carrion, offal, garbage, filth, birds, mammals, reptiles, fishes."

The first raven I ever saw seemed so somber and aloof I would no more have expected him to burst into play than a member of the Supreme Court to jump from her chair and break-dance on the chamber floor.

But play they do. In fact, of all the animals on the planet that romp, cavort, and frolic, ravens are among the most extremely playful—right up there with great apes, dolphins, and parrots. Hunter-gatherers in the Northern Hemisphere have known this for ages, says Osvath. One creation myth tells how a raven crafted humankind simply because he wanted a play partner.

Play occupies a big chunk of time for ravens, especially young ones. A few examples of their games: picking up twigs and flying with them, then drop-catching them in the air; hanging upside down by one foot while holding a toy or a piece of food, then switching the object from beak to foot and back again, time after time. One captive raven was observed tossing a rubber ball in the air again and again and catching it while lying on its back. A tipster in Arthur Cleveland Bent's *Life Histories of North American Jays, Crows, and Titmice* watched ravens sliding down the high bank of a river on rolling pebbles and crumbling clay, a dozen at a time, croaking loudly with apparent enjoyment. "This noise was heard over a mile before we paddled up to the birds, where we stopped to witness their amusement. The trees in the vicinity contained numbers of ravens aiding the sport with their cries of approval or taking their turns as the others became tired."

Dirk Van Vuren, a wildlife ecologist at the University of California,

Davis, witnessed common ravens on Santa Cruz Island performing elaborate aerobatic rolls like jet pilots. While in flight, one bird rolled on its long axis, first left, then right—nineteen times in a row. Another executed so-called Immelmann turns, named for the World War I aerial maneuver to reposition an aircraft for re-attack. From below, the pilot steers the plane back up past an enemy aircraft and then, just short of a stall, "yaws" the plane around, 360 degrees. This bird "rolled onto its back and then proceeded into a one-half inside loop, which it concluded gliding upright in the opposite direction," writes Van Vuren. One daredevil raven performed what sounds like the winning sequence of an Olympic figure skater: six half-rolls, two full rolls, and two double rolls.

As one of the most playful animals on the planet, ravens embody a mystery. Play is a strange behavior in any animal, says Osvath. "There are so many good reasons not to do it." It takes a lot of energy that could be used for other purposes—growing, for instance. It's also inherently risky. "If you're out in the wild, and everyone is playing, then no one is paying attention to any potential threat. A bird at play is particularly conspicuous to predators, which are never far away." The behavior seems at best extravagant; at worst, downright dangerous. "Out in the wild, playing must be very important," says Osvath. "Otherwise, why would you make yourself vulnerable in this way? It's a world of eagles, owls, even wolves out there. Distraction at the wrong moment could mean the end of you."

Why would a raven play? What possible benefits could there be to larking about? Do these birds play because they're intelligent, or are they intelligent because they play?

Osvath explores these and other questions at his research farm and aviary in southern Sweden, where he has studied ravens and raven cognition for a decade. When he offered to introduce me to his birds, I jumped at the chance. On a spring day, he picks me up in the ancient town of Lund, founded in the Viking age and home to Lund University. The gardens of Lund are blue with *Rysk blåstjärna*, which translates literally to "Russian blue stars." The morning air is warm, but there's a powerful wind blowing through the narrow, cobblestone streets, apparently typical in this part of Sweden, and a source of aerial play for the ravens that abound in the countryside. In town, it's rooks that struggle in the gusts, cawing loudly or carrying twigs in their bills. Around the university campus, treetops bloom with colonies of their big bulky nests.

Osvath tried experimenting with rooks. "They're smart, but they're difficult to work with when they grow up," he says. "They don't interact with humans, even if they know you." But it's odd how they're drawn to universities, he says. "You find them in Lund, in Oxford and Cambridge. And for some reason, you find them in Uppsala, the other very large, old university city in Sweden, in the very far north beyond their range, where you wouldn't normally find rooks. "I have a theory that they're very academic."

"But difficult to work with," I say.

"Yes," he says, "true of many academics."

Doing science with ravens, on the other hand, is easy. "They're so inquisitive and engaged," says Osvath. "If they find that a task is interesting enough, they'll line up to participate. They're fun to study, and they make me laugh." They're also as knotty and contradictory as any character from Shakespeare. They're highly intellectual birds, yet often silly in their antics. They have a strict, rigid social hierarchy and pecking order maintained through conflicts and fighting—they form coalitions and beat up on one another—and yet they continue to frolic together well into adulthood. They love to noodle around with objects yet seem deathly afraid of them—at least when they're new.

"If you put a chair in the aviary, for example," says Osvath, "the birds will sit and look at it for two weeks. And then, all of a sudden, they'll go for it, and it will take them like five minutes to completely dismember it. You never know what will scare them. You can think, 'Oh, when I bring in this new experimental device, they'll just freak out. But they don't. And then one day when you walk into the aviary, they completely lose it, and you can't understand what the problem is, until you realize, 'Oh, I'm wearing new gloves.' Those parts of their minds are very distant from the mammalian mind, and it's often difficult to understand them." Some years ago, raven expert Bernd Heinrich published a paper called "Why do ravens fear their food?," in which he speculated that if you eat carcasses, as ravens do, it's good to know that they're dead, so you observe an object thoroughly before making a move to dine on it. The ravens' intense fear of something new and unfamiliar, neophobia as it's called, makes their playfulness seem all the more extraordinary.

We're heading for the small village of Brunslöv, about twenty miles east of Lund, in the part of Sweden known as Skåne, a countryside of undulating fields spotted with old farms. Osvath was born in Lund and attended the university there. Now he and his wife, Helena, live in a 150-year-old farmhouse heavily populated with animals, including rheas and tinamous—two so-called palaeognaths, considered the most primitive of living birds.

When we arrive, a trio of rheas in an outdoor pen attached to the house poke, scratch, and frolic in the beach volleyball sand Osvath brought in to mimic the South American pampas they normally inhabit. Distantly related to ostriches and emus, the rheas are all appendages, known in the local Guarani language as *ñandú guazu*, or "big spider."

At the time of my visit, the kitchen was home to three elegant crested tinamous, charismatic birds of western Argentina, the only palaeognaths that fly, but perhaps best known for their eye-popping, highly glazed lime-green eggs, the shiniest and most conspicuous eggs in the bird world. Osvath was hoping to use the tinamous in his experiments testing cognitive mechanisms in palaeognaths—inhibition, working memory, learning, object permanence—but these particular birds have turned out to be too skittish to train. "They freak out if you get too close to them and start jumping in every direction," says Osvath. "They're afraid of their own food bowl." In their defense: Also in residence is a Maine coon cat, which sits on the kitchen table closely watching the tinamous, and a sight hound named Nimrod, who thinks he's a cat and stands on a table nearby. A hunting dog of champion stock, Nimrod will take off after a deer at forty miles per hour, but he's not a problem around the birds—in fact, he cleans the raven chick faces with his tongue. Once he brought Osvath a pair of baby pigeons. "Opened his mouth and there they were," says Osvath, "unharmed and so young they had no feathers." The coon cat, however, is a killer, so perhaps it's no wonder those tinamous are a little on edge (though Osvath insists they were skittish even in a large outdoor aviary).

In his training as a cognitive zoologist, Osvath focused first on great apes, gaining fame when he published studies in 2008 hailed as the strongest evidence yet that a nonhuman animal could plan ahead. Following this breakthrough paper was his rather comical report of the planning behavior of that male chimp named Santino in a Swedish zoo, who would collect stones and other missiles and hide them in strategic spots so that later, when the mood hit, he could use them to hurl at gawping visitors.

Osvath turned his attention to ravens in part, he says, because great apes are so like humans that for him they were "a bit boring." "Sometimes—as a scientist I should perhaps not say this—but sometimes, ravens seem to be cleverer, at least in some areas."

Among the recent findings about the intelligence of these birds: A 2019 study by Osvath and his PhD student Katarzyna Bobrowicz showed that ravens take only half as long as humans to select an object out of an array of objects based upon a single glance at the choices. This study suggests that the raven's cognitive processing of visual input is faster than ours; its visual system picks up more information per time unit than a mammal's does. Such speedy perception might affect the general speed of cognitive processing in more complex challenges, as well. "They outperform apes at many tasks, which is so amazing because they're *dinosaurs*," says Osvath. "And interacting with them is quite different from other birds, quite extraordinary. In this way, they're more like dog pups compared to rooks or jackdaws."

Corvids and apes are separated by 320 million years of evolution, but in terms of their complex cognition, they're surprisingly similar—which for Osvath raises some fascinating questions. Did birds and primates evolve their abilities independently? Or did they build on abilities that were present in their common ancestors hundreds of millions of years ago? He keeps the rheas and tinamous for his study of the deep evolutionary history of the bird lineage, comparing them with alligators, which he keeps at a facility an hour down the road from the farm. He considers alligators highly intelligent reptiles, "inquisitive, explorative, socially nice," he says. "That is, if they know you quite well. They could bite very hard if they wanted to. But instead they nibble gently on your boot if they want to tell you off and you're not quick enough on your feet." Crocodilians are important for understanding the evolution of neural systems in animals because they're the nearest relatives of modern birds and share an ancestor with birds and mammals. "I'm hoping that an evolutionary perspective proves fruitful for understanding both complex cognition and complex play."

Osvath has studied primates and rooks and other corvids through this lens of evolution. But his passion is for ravens. In 2008, he and his wife Helena began building their spacious aviary, complete with an enclosed observation room that has a large plate glass window where Osvath can watch the ravens—and they can watch *him*. Then the pair raised a group of raven chicks from hatching. For weeks before the aviary was finished, they had twelve raven chicks in their bedroom. "The chicks woke hungry every morning at five, so we'd feed them," says Osvath. "That's how you form strong bonds." Now the ravens trust them both almost completely and sit comfortably on their arms or shoulders. Osvath's graduate students go through a bit of hazing to earn the birds' trust. "The ravens often test them," he says. "One time, they gave some really hard pecks on the head to a postdoc researcher who had just arrived, which led to bloodshed. But thereafter, when you've proven your place, they are more than happy to work with you."

The aviary is currently devoted to six of the "grim, ungainly" birds, two pairs, Siden and Juno, and Rickard and None, as well as two females, Tosta and Embla. "Some ridiculous names," says Osvath. "The dominant male is Siden, which is Swedish for 'silk.' There's nothing silky about him. But we called him Siden because we tied a little silk string around his leg when we first transported him." And None? "When we named the chicks for the experiment protocol, she was the only one without an identifying ring, so we just wrote 'None.'" There's a seventh raven on the premises, but she's wild. "She doesn't understand that she's wild, so she hangs around," says Osvath. "And when we do experiments, we open the aviary, and she just comes in, participates, and then at the end, walks out. In the ethics section of grant applications, it's difficult to explain: She's wild, but her participation is completely voluntary."

Among Osvath's more striking studies is one that appears to propel ravens into that exclusive clutch of animals—apes and scrub jays—that may use past experience to plan for the future. Osvath and a graduate student, Can Kabadayi, taught five ravens how to use a special tool—a stone of a specific weight and shape—to open a puzzle-like box containing a high-value treat. Though ravens are not tool users by nature, the birds were quick studies, requiring only one lesson to learn the trick—"a single observation of tool use by a non-tool-using species!" Osvath exclaims. When the experimenters took the box away and presented the ravens with a choice of objects, only one of which was the functional tool that could open the box, the birds picked the object that would be most useful to them if that treat-laden puzzle box reappeared in the future. They could wait as long as seventeen hours and would do so even when they were offered an inferior treat immediately; despite the temptation, the birds didn't bite on the lesser alternative. Instead, they chose the special tool and saved it to secure the later, tastier reward.

The experimenters also taught the birds how to trade that tool for a token, a bottle cap, that would earn them an even better reward. On tests in which they had to barter for the superior treat, the birds outperformed apes and even young children.

In both tests, ravens showed abilities of self-control, reasoning, and flexible planning for the future equal to a great ape's.

So clever are these birds that they occasionally outwit their experimenters. Osvath ran into trouble with one female in this study who invented a completely different way of opening the puzzle box, using bits of bark she found scattered about. She had to sit out the experiment on the sidelines.

The floor of the raven aviary is littered with toys—sticks, bones, balls, boots, a hay bale, tubs for bathing—and perches, stable and swinging, which the birds use like balance beams or trapezes. Often, the ravens will hang upside down by their feet from a branch, with wings outstretched, then let go—first with one foot, then with the other. "As far as I can tell, the game is about not extending your wings until it's absolutely necessary," says Osvath. "Or they'll hang upside down with their beak, holding an object in one foot, letting go of the object, then letting themselves go and somehow trying to catch the thing, never succeeding, but then doing it all over again." Other ravens sitting around watching will suddenly join the game.

Play is not an easy behavior to pin down in any animal. We tend to look for behaviors like our own childish frolicking—kittens tussling with a ball of yarn, calves cavorting, puppies thrashing a stuffed toy. But play in species more distantly related to us is more enigmatic. "It's super obvious in ravens because they play like we play," says Osvath, "but we might miss it in some species that play in a way we can't relate to. We think we know it when we see it, especially in its extreme form, but it's very difficult to define."

In fact, play is most often defined by what it is not: It is behavior that is not purposeful, serves no obvious adaptive function, does nothing to enhance an animal's chance of survival and reproduction—nothing apparent, at least.

Not long ago, Gordon Burghardt, an evolutionary psychologist at the University of Tennessee, Knoxville, took on the problem of crafting a more positive definition of play and came up with a set of five criteria for recognizing play in any species. First, it's behavior that's not fully functional, although it may resemble behavior that's useful in another context. For instance, a puppy thrashing a toy mimics a wolf's takedown of prey, but the puppy doesn't end up eating the toy (well, not usually, although my pitbull-lab mix has been known to). Play is exaggerated, awkward, and repetitive—like all those excessive flourishes in flight of the ravens on Santa Cruz Island—and often inappropriate. It's also spontaneous, voluntary, intentional, pleasurable, and rewarding. It's only initiated when an animal is adequately fed, healthy, and not under stress. In sum, says Burghardt, "Play is repeated, seemingly non-functional behavior different from more adaptive versions . . . and initiated when the animal is in a relaxed, unstimulating or low stress setting." In other words, it's a kind of behavioral frill—but far from a frivolous one.

Burghardt admits that this thumbnail description doesn't capture all the nuances of play, but it works to identify the behavior in animals and situations where it was previously ignored or dismissed or just not taken seriously.

It has been obvious to naturalists for decades, probably centuries, that animals play. But it was thought that the activity was limited to mammals—humans, horses, chimps, cats and dogs, otters, dolphins. The idea that a wild bird, or any wild nonmammalian creature, for that matter, might have the luxury to pursue fun or entertainment or activity beyond its basic needs seemed hard to swallow. But in recent years, Burghardt and others have found play popping up in unexpected places in the animal world. Octopuses toy with Legos and toss balls with their jets. Saltwater crocodiles whack at a tethered ball. At the National Zoo in Washington, DC, monitor lizards play games of keep-away, and

Komodo dragons engage in tug-of-war with their keepers and have been known to pluck a keeper's little notebook from a pocket and parade around with notebook in mouth, like a dog trotting around with a shoe— but in slow motion. Poison dart frogs play rough-and-tumble with each other. Even fish have been known to leapfrog and bat balls and balance twigs on their noses. Burghardt watched captive soft-shelled turtles play with a basketball, nosing it, batting it, chasing it all around their enclosure.

And birds?

All variety of birds throw things. Rainbow bee-eaters, warblers, and pelicans throw pebbles, and neotropic cormorants and green herons toss sticks, leaves, pods, and fish in the air. A number of birds seem to enjoy a good ride. Adélie penguins surf on small ice floes along a tide run like otters on a water slide. Rainbow lorikeets swing from tree limbs. An Anna's hummingbird was seen floating down the stream of a water hose, then returning to the top to repeat the drift, over and over. Pied currawongs play tug-of-war with sticks and also a version of king of the castle, with one bird occupying a big perch while others try to knock it off. The winner assumes the position, only to be knocked aside by another bird. Andrew Skeoch has seen a group of choughs lying beside one another in a row, on their backs, legs in the air, passing a stick back and forth with their feet. He has witnessed the behavior repeatedly in this particular group but never in other chough groups, so he believes it's a socially learned game. In their wonderful book *Thinking Like a Parrot*, Alan Bond and Judy Diamond write that kakas "engage in wild bouts of rough-and-tumble play, jump on each other's backs, roll over and wiggle their feet in the air, jump up and down on each other's stomachs while flapping their wings."

And parrots generally? Why eat that flower sitting upright when you can roll yourself over a branch, hang upside down, and nibble on it from below?

Gulls drop-catch clams and other objects, letting them fall from the air and then swooping down to snatch them again midair. The behavior is more common in a stronger wind, suggesting the gulls might play the game more when the task is more challenging. Turquoise-browed motmots of Costa Rica bounce on their food. If a motmot has recently eaten, it "might go by a piece of food, apparently ignoring it, and then whirl and pounce on it," writes researcher Susan M. Smith, then at the Universidad Nacional de Costa Rica. Or it "would look over its shoulder at its own tail, then whirl around in evident pursuit, often making four or five complete revolutions before stopping." On one occasion, Smith saw a bird go around four times in one direction, stop, then pounce suddenly in the other direction. "I saw tail chasing nine times involving five birds. No bird succeeded in touching its tail during this behavior."

Now that's "exaggerated, awkward, and repetitive."

Arabian babblers, especially young ones, play a variety of games—sometimes for several hours each day—wrestling, cockfighting (where two babblers jump head-on at each other, each trying to make the other lose balance), tug-of-war, and king of the hill.

Birds play. Not all species play as often or in such complex ways as the most extreme species, says Osvath. "But if I had to put money on it," he says, "I would bet that the majority of birds play. I've witnessed indisputable play in rheas and emus"—species considered "primitive." Even birds with no general penchant for playfulness occasionally get frisky. I was charmed by Joe Hutto's account of disport among the wild turkeys he came to know so well in the flatwoods of Florida: "Very young wild turkeys are much too serious and cautious to engage in any activity that resembles play," he writes, "but as they grow in size and begin to mature, they allow themselves a little latitude and show certain exuberant behaviors that resemble play—spontaneous, short, flight-assisted jumps, ducking and dodging as if some imaginary adversary exists."

Scientists recognize three main types of play, and ravens revel in all of them. There's locomotor play performed using just the body—running jumping, kicking, twirling. Roughly half of all twenty-seven orders of birds engage in this sort of play, from penguins to hummingbirds. Object play is defined as the repeated manipulation of "inappropriate" items—picking up objects, leaves, sticks, stones, tossing or dropping them, ripping them, bouncing up and down on them—and is practiced by six orders of birds, from gulls, raptors, and owls, to woodpeckers, songbirds, and parrots. As far as we know, it is mainly these last three orders that engage in the rarest and most sophisticated form of play, social, described as playful engagement with others in a safe space—wrestling, chasing, mock fighting. What makes this kind of rough-and-tumble activity "play" is that it holds to a set of rules: play fair, don't inflict pain, and above all, take turns, a fundamental social skill. Sometimes, an animal will combine different kinds of play into mixed patterns, called "super-play" by researcher Sergio Pellis. Osvath has watched ravens superplaying, hanging upside down while manipulating small objects.

All of this makes play sound common in birds, but in fact it's documented in only about 1 percent of the ten-thousand-plus species. Osvath suspects that's because only this tiny percentage has been thoroughly studied.

There are three prerequisites for complex play, and the raven dossier lists them all. The behavior occurs in birds that have a brain big for their body size (ravens have among the largest relative brain size of any bird), a long juvenile period when youngsters hang around with their parents (ravens spend at least five months with their family), and a habit of exploiting all available food resources (ravens specialize in being food generalists). Most birds that engage in social play have a complex social system. Ducks do something called coordinated loafing, hanging out together. But real social play exists mostly among birds that live in complex societies where alliances are fluid and constantly shifting, like ravens and Australian magpies.

When we enter the observation room next to Osvath's aviary, the ravens eye me warily. We can't go in because the birds will attack me. It's breeding time, and the pairs, Siden and Juno, and Rickard and None, have nests full of chicks. Ravens are fierce defenders of their young and peerless providers, with highly efficient hunting skills. Once, says Osvath, a male was perched on his hand. "He flew up in the air, looked down, and saw a rat. In a split second, he dropped on the rat and killed it. He opened it up, removed the intestines, and chop, chop, chopped the meat into little bits and fed it, still warm, to his chicks."

We watch from a distance while Siden moves a huge hunk of horse-meat to a spot next to a bathtub, then finds a black rubber boot and drags it over to cover the meat. A few flicks of straw, and the meat is well concealed. At least to our eyes. Siden's mate, Juno, immediately swoops down from the nest, plunders the hidden stash, and takes it back to her nestlings.

This secreting of meat may look like a game of hide-and-go-seek, but it's really about caching food. Ravens are master cachers—but they don't start out this way. Studies show that juveniles learn the art of a good hiding place through trial and error. It can be a steep learning curve. Osvath says a neighbor once found a large lump of meat cached in the mane of his horse. Researcher Raoul Schwing, who occasionally helps with raven studies at an Austrian aviary, recounts even-more-inappropriate hiding places. "You'll be cleaning the aviary and you lean forward so there's a bit of plumber's crack showing, and suddenly you have a piece of dead meat stuffed down your pants and a young raven hopping around behind you, 'Oh look, I can cache something!'"

Studies show that adult ravens know when other ravens are spying on them while they cache, and will use a variety of strategies to protect their caches from pilfering: delaying caching until competitors have left, staying away from caches they've already made if a competitor is present, using obstacles as visual barriers, and seeking caching locations where other birds can't watch—behind trees, rocks, or in the case of the aviary, bathtubs or piles of branches or old boots. In an ingenious experiment, Thomas Bugnyar of the University of Vienna and his colleagues found that ravens will guard their caches against discovery in response to the sound of another raven in an adjacent room—but only when a peephole between the rooms is open and not when it's closed. This suggests that ravens are more than mere "behavior-readers" and can take the perspective of another bird, holding in mind what another creature can or cannot see—which some scientists consider a critical component of the sophisticated cognitive ability known as theory of mind.

A raven chick will eat as much as a pound and a half of meat per day. "Last year, we got a whole slaughtered horse in April for the birds in just these two nests," says Osvath. "It was all eaten by the end of July. Imagine how these birds have to work in the wild. When the chicks are young, the male must feed his partner, all the chicks, and himself." Which goes a long way toward explaining why, at the moment, the raven parents are not feeling particularly playful.

But just a few weeks ago, all six birds were playing in the snow, "which they absolutely love," says Osvath. "They do all the classical things, sliding down a hill, going back up, sliding down again. Sometimes on their backs with a stick in their feet. You can throw snowballs at them, and they all line up and try to catch them, jumping as high as possible. Sometimes one raven will walk over to another and just grab a leg and yank it. Then the other will grab a leg, too, and they'll both lose balance and fall over in the snow, *boom!* holding each other's legs, and sometimes holding an object in the other foot as well. It looks absolutely crazy."

The ravens also love to bathe—not to get clean, or to warm up or cool down, but to frolic and splash in water. Bernd Heinrich established this in a series of experiments aimed at determining whether the birds chose to bathe when they were dirty or when they needed to change their body temperature. To dirty the birds, he sprayed them with honey and cow dung, but this made no difference in their rate of bathing. And it didn't matter whether the temperature was a frigid minus-45 degrees Fahrenheit in the winter or a blistering 90 degrees in the summer. "Temperature just wasn't a variable," says Heinrich. They bathed when they felt like bathing, for the sheer fun of it.

"The first bath after Sweden's deep winter is especially fun to watch," says Osvath. The ravens flock to the tubs and pools he fills around the aviary. They stand on the brim and dive underwater again and again, cawing and splashing.

Ravens are renowned for their ability to invent all variety of games. "If you go to different raven facilities, you'll see the birds playing different games, which individuals have come up with themselves," says Osvath. "Introduce them to a new game, and they learn the rules quickly." In the 1960s, Eberhard Gwinner, the inspired zoologist who revealed the internal calendar in migrating birds, turned his attention to play behavior in corvids—ravens, carrion crows, and rooks—noting how individual ravens devised their own new games, which other ravens imitated and continued playing for years even after the inventor was gone—in what, he said, amounted to a culture of play.

Give a raven a jar or an empty flowerpot, and it will put its head inside and listen to the sound of its own voice, says Osvath. "It's the same sort of sensory play you see in kids—it's fun to talk into something that alters your voice." Raven nestlings will sit in a corner and gurgle away to themselves. One male raven in Osvath's care had injured his wing when he was young. So as not to strain the wing, the bird was housed in a smaller aviary near an electric fence enclosing the horses on the farm. The fence clicks, and the horses back away from it when they hear the sound. The male raven learned to imitate the click just to get a rise from the horses. "He probably realized, 'Cool, by doing this, I can move tons of meat!'" says Osvath. "This became his trademark self-advertising display call: He fluffs himself up to make himself look bigger and out comes this little 'click!'"

Some birds repeatedly drop objects and listen for the sounds they make. Galahs in New South Wales were seen dropping pebbles on the metal roof of a country house. Male spotted bowerbirds have been known to pick up shells from a heap and drop them on the pile, two or three times, apparently just to hear the little tinkle. In her book *Avian Play*, Millicent Ficken notes that young garden warblers have been observed doing this, too. "One bird dropped a stone accidentally in a glass making a ringing sound; all the other birds showed great interest, and many began dropping stones in a dish."

This kind of vocal or acoustic play doesn't fit into the three formal play categories; perhaps it's a fourth. Some scientists suggest that subsong—the soft, random, almost-but-not-quite-adult sounds that young songbirds produce—may be a form of vocal play. Studies by Lauren Riters of the University of Wisconsin and others show that song practice and so-called undirected singing of any kind outside the breeding season and at any age is intrinsically rewarding for a bird, involving those same opioid pathways active in experiencing pleasure. If that's the case, if singing outside the breeding context is a form of play for the 4,500-plus species of oscine songbirds, then the estimate of play in only 1 percent of bird species is off by a lot.

The first year the wild raven turned up around his aviary, Osvath saw her playing with a red kite of roughly the same age. The two birds would hang out together, sitting next to each other and occasionally doing aerial play. Sometimes it was the red kite that initiated the frolicking, sometimes the raven.

Play between species is not all that unusual, but it's often surprising. YouTube is full of entertaining clips—of a dog and a pony horsing around; a kitten tussling with a fawn; a magpie rollicking with a puppy; a German shepherd nuzzling a macaw; a Chihuahua cuddling with a chick; a kangaroo petting a dog enthusiastically as if to say, "Who's a good dog? You are! You're a good dog!"; and, perhaps most astonishing, a barn owl cavorting with a cat and huskies roughhousing with polar bears, even taking hugs from them.

"Animals quickly understand when another animal is playing," says Osvath, even if it's a different species. You can see it in humans, apes, dogs, cats, birds. There must be a conserved sign, a characteristic behavior, that all—or at least most—vertebrates understand." Some animals have distinct play signals, like a play bow in a dog, front legs spread, rump high and tail up and wagging, or the play face of chimps, a hybrid of grimace and smile, or the way baboons lean over and look between their legs. But there are other signs, says Osvath, such as an extra-bouncy stride or exaggerated movements. Osvath once played with a brown bear at a zoo, running to one end of its enclosure, touching nose to glass, then running to the other end. "The bear immediately got the game," he says. "Chimpanzees will do this, too; they catch on quickly." Videos on the internet show penguins in underwater zoo displays swimming back and forth in their tanks, playing beak-to-nose tag with human visitors. Birds generally invite play by cocking their heads, rolling over onto their backs, sidling up to a playmate, or just generally bouncing around.

Osvath has found that raven play starts young, even before nestlings have fledged. "We could observe them in the nest at around forty days after they hatched, when they were big enough to be above the brim, and they were playing like crazy—yanking, biting, pecking at bits of nest material—both alone and together." Fully a third of the young birds' time in the nest was spent playing, second only to hours spent sleeping, and three times more than time spent in flight training. "Later, as juveniles, that's more or less all they do," says Osvath, "play and eat." The play mood among nestlings appeared to be infectious. When one bird started to play, another would join in.

Osvath later discovered that this play contagion wasn't a simple matter of so-called behavioral synchronization, when birds align their behavior and do the same thing side by side. If one raven was engaging in object play, another might start doing locomotor play, and still others, social play. "It's the play *mood* that seemed to be contagious," says Osvath. Positive emotional contagion is also found in other animal species— chimps and rats, for instance. Scientists have lately found that negative emotions in ravens can be contagious, too. When a raven sees allies struggle with a task that denies them a treat or sees their disappointment in response to unappetizing food—but not the food itself—its own interest in food diminishes. This kind of emotional contagion, whether positive or negative, is considered a building block of empathy.

Though we can't see inside Siden and Juno's nest, it's likely the chicks there are hard at play, pulling up pieces from the bottom of their nest and frisking with one another in the tight space—impish antics that flutter in the face of the bird's somber stereotype. Even as we speak—and even in breeding season—one of the male adults, Rickard, sidles back and forth down a log suspended by chains, setting in motion a brisk back-and-forth swing.

Osvath is curious about why ravens have joined a fairly exclusive club of exceptional players in the animal world. Just a few animal groups stand out for their complex, innovative play behaviors—great apes, dolphins, parrots, and corvids. This kind of play seems to pop up only sporadically on the phylogenetic tree of animals and on very distant limbs. Of the thirty or so phyla in the animal kingdom, only three seem to contain species that play. And within those three, not all lineages appear to have playful species. On the bird branch of the tree, the top players—parrots and corvids—are only distantly related, separated by ninety-two million years of evolution.

That play—and especially complex play—is spotty among animals suggests that it evolved independently multiple times and may be important in some way. But its purpose has been a mystery, the hardest nut to crack, says Osvath. "We still don't know why they do it."

The traditional view is that play has a single function—to hone vital life skills needed in adulthood, like hunting or fighting. "The first thing people think of is 'Oh, it's practice,'" says Osvath. The idea dates back to Karl Groos, who published a book in 1898 called *The Play of Animals*. "The play of young animals serves to fit them for the tasks of later life," writes Groos. "Play is surely not the child of work," but rather "work is the child of play," the training ground in the struggle of life.

There's not a lot of experimental evidence supporting a link between youthful playing and adult expertise, but it makes sense. Osvath has watched young ravens toy with nearly every new object they meet: twigs, stones, inedible berries, bottle caps, bits of seashells and glass. This kind of early play may help young birds distinguish safe objects from dangerous ones. And manipulating objects appears to help juvenile ravens develop their caching skills—to figure out how best to camouflage their stashes and use visual barriers to conceal their efforts from the prying eyes of pilferers. Thomas Bugnyar and his colleagues also found that playful object caching, also known as "stupid" caching—caching inedible play items such as small stones, twigs, colorful plastic objects—gives ravens a chance to evaluate the pilfering skills of other ravens without risk of losing valuable food, information they can later use in genuine food caching.

It's also easy to see how complex social play early in life might lead to smoother social relations later on, promoting social bonds, establishing dominance hierarchies, and teaching animals how to interact with others in positive ways.

Likewise, it makes sense that locomotor play—diving, swooping, etc.—might be useful in learning to avoid predators. But it would surely be a stretch to claim that a raven's aerial looping or dropping like a rock from the sky would be useful in this context. "They're so agile in the air that no eagle could ever harm them," says Osvath. "And their locomotor play is often about challenging themselves and then increasing the difficulty for no other reason than to make the game harder. It's self-handicapping, which you see in many animals we view as clever." For instance, they'll hold an object with one foot and drag it in a very awkward way and then carry it out to the end of a long, thin branch and try to maintain their balance. "That's typical raven play," says Osvath, "doing things they would never do normally, things that don't make sense in any biologically relevant context."

Gordon Burghardt suggests that playful actions like the bubble blowing of dolphins and pirouetting of chimpanzees and the way some monkeys and apes run with their eyes closed may have originated in practice or training for a particular function, but then over evolutionary time, these actions took a gamboling detour, says Burghardt, and "became divorced from that underlying functional system." Osvath agrees that certain kinds of play may have evolved as a by-product of training skills and then started to serve a different function. "And that function may be different for different animals," he says. Or it could have gone the other way.

For instance, playful object-oriented explorations in the early ancestors of corvids may have led to the innovation of caching.

Play may be training for the unexpected. Half a century ago, Eberhard Gwinner suggested that play shapes the remarkable behavioral flexibility of ravens, allowing them to develop a more "diverse and responsive repertory" of behaviors in the face of environmental challenges. This idea was recently revived by Marc Bekoff and his colleagues, who propose that "play functions to increase the versatility of movements and the ability to recover from sudden shocks . . . and stressful situations. This is why animals actively seek and create unexpected situations in play and actively put themselves into disadvantageous positions and situations." Play allows an animal to explore options in a safe setting.

It also may reduce stress. In some young animals, low levels of the stress hormone corticosterone are linked with high amounts of play. Unstressed animals are perhaps more likely to play. But there's also the possibility that play relieves stress, either in the moment or in response to stressful events in the future. Some kinds of play fighting, for instance, include elements of real fighting and activate the same neurochemical pathways in a bird's brain as the fight-or-flight response. By creating little peaks of mild stress in safe circumstances, play fighting might alter a bird's sensitivity to stress so the next time the bird encounters a truly stressful challenge, it's not as traumatized and recovers more quickly.

One way to find out what play is good for, says Osvath, is to take it away from a raven and see how it does. "But this would be nearly impossible to accomplish because you would have to physically restrain the bird, which would not only be unethical; it would cause extreme stress, a confounding factor."

Laboratory rats deprived of play don't develop normal brains. Scientists discovered this by housing juvenile rats with adults, which inhibits their play impulses. The young rats, which had all of the other kinds of social interactions—touching, sniffling, etc.—but no play, did not develop a normal prefrontal cortex. This, then, is another possible vital purpose of play in all animals: the healthy development of the brain.

Osvath has what he calls a "very shaky hypothesis" related to this. "Not even a hypothesis, just a thought," he says, but it's an intriguing one. Maybe play, especially for adult ravens, has a regenerative function for the brain. We know that the birth of new neurons—neurogenesis—occurs in birds that cache their food in the winter, especially in the hippocampus, the part of the brain responsible for learning, memory, and regulation of mood. (Evidence suggests that humans also have the ability to grow new neurons in the hippocampus throughout life.) "Maybe it's like sleep, another mystery," he muses. "You need it to regenerate and consolidate memories."

Of course, there's a proximate explanation for why ravens play, says Osvath: Because it's fun. "We scientists are not supposed to say that, but almost every one of us—between the papers, so to speak—would agree that the animals we see playing are having fun, and that fun can be its own powerful reward." In a recent study on object play in crows, ravens, and jackdaws, Osvath and several colleagues from Oxford University and the Max Planck Institute for Ornithology found that ravens offered the choice of food or play objects most often chose the latter. "The object play was so important they could say no to food," says Osvath.

Research suggests that birds have the potential to experience fun in the same way mammals do. In an essay called "Do birds have the capacity for fun?", cognitive scientists and corvid experts Nathan Emery and Nicola Clayton cautiously suggest that the answer is yes. In most animals, play seems to be a powerful trigger for the release of dopamine, which is active in the reward system in the brain, and also for endogenous opioids, which are essential for sensations of pleasure. Although we don't yet know whether similar processes occur in birds, we do know they possess the same neural underpinnings of fun and reward. "Dopamine appears to play an essential role in reward in birds and is found in analogous brain regions," write Emery and Clayton, "suggesting that it also controls the search for reward-inducing stimuli in birds." We also know that social play and undirected singing in birds is tightly coupled to the activity of endogenous opioids in the brain, which may flood those areas of the brain linked with sensations of pleasure and reward.

The reward system is central to everything an animal does, says Osvath. It's why we eat and why we have sex. It's why some of us watch birds and why some of us study them. "A lot of us do things for no particular purpose," says Osvath, "science, culture, writing books—we do these things mainly because we like to do them. It's very much associated with our exploratory or play behavior. It's completely useless to study birds or go to the moon or investigate any field that does not have practical applications. It's getting basic knowledge, and that's very interesting to us and we learn a lot from it, but it's actually just play."

Chapter Eight

CLOWNS OF THE MOUNTAINS

Walking into the kea aviary at the Messerli Research Institute feels a lot like being the favorite aunt arriving at a loud and boisterous family reunion. The birds beeline to greet you like swarming children, waddling up at full pelt, squawking, clustering at your feet, tugging at your pockets as if to say, "What did you bring me? What did you bring me?"

There are two labs that study kea, one established only recently in New Zealand by Alex Taylor, a researcher at the University of Auckland, and the one I'm visiting, perched halfway around the world from New Zealand, in the beautiful green rolling countryside of lower Austria. This one, launched thirty years ago by cognitive biologist Ludwig Huber, is a collaboration between two universities in Vienna and is now located at the Haidlhof Research Station just outside the town of Bad Vöslau. It includes laboratory aviaries for ravens, led by Thomas Bugnyar of the University of Vienna, and for kea, led by Raoul Schwing of the University of Veterinary Medicine, Vienna. Schwing runs the kea lab with the help of his manager, Amelia Wein, who has offered to give me a tour of the aviary and let me watch the afternoon kea feeding. It's April, right in the middle of the breeding season for the kea, just as it is for the ravens, but Wein tells me that it's perfectly fine to enter the kea aviary and meet the birds. However, she warns, "Leave everything you can outside the aviary—phone, watch, earrings, everything."

Now I see why. The kea amble over one by one with that curious parrot wobble. They explore every inch of my body, plucking at my jeans, pulling at my shirt, my socks, my shoes, examining each new item with singular curiosity, nudging things one at a time with their long, curved beaks the way a toddler might poke at something new with an index finger. In fact, they remind me of nothing so much as my own daughters when they were infants grabbing on to my hair or earrings and shrieking with delight as they pulled hard and then tried to untangle their little hands. Suddenly the birds descend on my sneakers and get busy with my shoelaces, tugging at them until both sets are undone and the ends start to unravel. Wein laughs. I look over at her, and she is literally coated in kea—two on her shoulders and one on her head, which has handily removed her plastic barrette and is now snipping away at her hair as though it had barbered in a former life. A small company at her feet gnaw away at her shoes. She bumps one bird off her shoulder. "*No!* I *like* this shirt."

All of Wein's shirts are full of little holes from kea beaks, she tells me. One of the birds' favorite activities is enlarging the pokes and perforations. They also love to take your finger or your earlobe or your Achilles tendon in their beaks and squeeze it not so tenderly. "If their beaks were like those of other parrots," says Wein, "we would all be missing fingers." But a kea beak is built not to crush nuts or remove seeds from their casings like most parrot beaks, but more like a hook, the better to dig up plants by the roots—lilies, daisies, speargrass—and to flip over rocks or logs and to explore, it would seem, anything in their paths—garbage cans, solar panels, cars, tents, shoelaces.

Kea are large, robust parrots, a gorgeous olive green on their backs, and underwings ablaze with a fiery reddish orange. The most extravagantly playful of all birds, they're called "clowns of the mountains" in their native region, the Southern Alps of New Zealand, and are renowned for their big brains and sophisticated intelligence—and also for being mischievous and destructive.

The kea and raven aviaries at Haidlhof are the setting for intense research on cognition and play in both species. As you approach the station, the hills are alive with the sound of—well, not music, but a cacophony of happy screechings and whoopings. The kea aviary houses twenty-eight birds, half adults, half juveniles, all descended from birds that survived the outbreak of West Nile virus that struck hard in Europe some years ago, and all named by Schwing and his colleagues. There's Plume and Papu and Pancake, Coco, Frowin, Pick, John, Lilly and Willy, and the certified genius of the lot, Kermit. Since he joined the lab in 2008, Schwing has been exploring the kea's trademark intelligence, fearlessness, curiosity, playfulness, and unusual social behavior, a constellation of traits that make this parrot truly unique in the bird world.

"I always say that kea are weird birds from a group of really weird birds," says Schwing. "Parrots in general are odd, but kea just take that to another level—in their evolution, their natural history, their behavior." Kea are the world's only alpine parrot that experiences mountainous winter conditions. Along with the kaka and kakapo, they form a distinct family on an ancient branch that diverged from other parrots, such as budgerigars and cockatoos, around fifty-five million to eighty million years ago. Some of the kea's features are strangely falcon-like. In flight they look more like a bird of prey than a parrot, and for a parrot, their syrinx—their vocal organ—is rather primitive, more like a falcon's. They have complex vocalizations, but they are not able to imitate human speech the way a cockatoo or an African grey parrot can. But so cheeky and humanlike are the kea, you can't help wondering what they would say if they could talk.

Ravens and kea are considered the two most playful birds on the planet, but they couldn't be more different in their demeanor. "The initial idea behind this station was to compare ravens and kea," says Schwing, "two fairly similar-size birds from the two different groups that are considered the smartest, corvids and parrots." Both birds live in large, complex social groups. Both have a flexible and generalist foraging style and are good at problem solving.

But there the similarities seem to end. For one thing, kea live together in relative harmony, while ravens are highly territorial and have a society with a strict hierarchy settled by fighting. Then there's the profound difference in their attitude to novelty. Ravens, as we know, are skittish around new things. Kea are drawn to them like ants to sugar. They have boundless curiosity and no fear. Whereas a raven will stare at an unfamiliar object for weeks before touching it, any new thing sparks instant investigation by a kea: "Hmm, what's this? What can I do with this?"

"Their huge love for novel things is a bit of a mystery," says Alex Taylor. "These birds move between subalpine forests and alpine areas. There's not much *new* stuff for them to love up there. That they have evolved this thirst for novelty is really strange. But it ties in beautifully with their levels of play."

Jorg Massen, Schwing's colleague who studied ravens with the Bugnyar lab, didn't fully buy the neophilia-neophobia distinction between kea and ravens, so he conducted an informal test. He sat down next to the aviary for the ravens, birds he had hand-raised himself or worked with closely, so they knew him well. They expected good things from him and came over to see what he was doing. He pulled out his bike light, and the instant he turned it on, the birds fled. Then he went over to the kea aviary. The kea saw him walk by their aviary every day, so he was familiar, and when he sat down, they ignored him. Until he turned on the light. He was instantly cluttered with a chaos of kea.

This dichotomy affects how the different labs at Haidlhof train students to work with the two birds, says Schwing. With the ravens, students are asked to stand outside the aviary for something like two to three weeks to give the ravens a chance to get to know them before they're allowed to interact with the birds in any way. "With kea, we do the opposite," says Schwing: "We train our students by sending them into the aviary with the caretakers to help with the cleaning, so the kea get to know them in a routine setting. Because once we start testing, we don't want the students to be so novel and interesting to the kea that they ignore what they're supposed to be doing and just go play with this new person they've never seen before," he says.

"I would argue that ravens are as bold as kea—but only with things and people they know. Kea go for everything new. This difference—the ravens being drawn to what they know, and the kea, to what they don't know, to new and novel things—makes a huge difference in how they behave and actually means that we must be very careful in designing our experiments and interpreting behaviors. If I'm always cautious, then part of my brain is always focused on that, on alertness. And that's not true for the kea."

Why has a bird that lives in the mountains evolved a mind that craves novelty and a desire to play all the time? Do kea play for the same reasons other birds do?

When Schwing joins us in the aviary, it's clear *he's* the favorite uncle at the family reunion. The birds crowd around him, squealing the long, loud, high pitched *keee-ee-aa-aa* that gave this bird its Maori name. It's a "contact call," what one kea uses to greet another.

To say these birds are vocal is putting it mildly. They are positively raucous, chattering constantly, mewing, screeching, trilling, and warbling to one another—and to whoever's in the aviary or just outside it. When I listen to my digital recordings of conversations with Schwing and Wein, the kea vocal commotion is so loud I can barely make out the human voices.

Schwing has been studying these birds for a dozen years, including more than four years investigating them in their native habitat in New Zealand's Southern Alps. "If you've ever wondered how green the mountains of New Zealand are"—Schwing points to a bird tweaking the rubber on his sneakers—"this is how green they are. You watch the kea with binoculars flying over the forest, and they land in the bushes, and they're just *gone*. It's incredible how well they blend in."

Kea have great camouflage. What they don't have is behaviors that might protect them from predators. In fact, quite the opposite. Their behavior seems designed to point a bright arrow at their big green bodies and to distract them from vigilance of any kind. This kind of bold behavior may be possible because kea have few natural predators in their native habitat. Two large birds of prey live there, the Australasian harrier and the New Zealand falcon—"pound for pound considered one of the most aggressive birds of prey in the world," says Schwing—but neither harrier nor falcon has ever been seen killing a kea. "In fact, the New Zealand falcon is half the weight of the kea," Schwing points out, "and the kea actually harass the falcons, dive-bombing them, all the while making play calls as if to say, 'Come join the *fun!*'"

In the past, it was another story. The huge Eyles's harrier, with a reported wingspan of up to eight feet, and the giant Haast's eagle, considered one of the largest, heaviest eagles ever described, dominated the island. Both went extinct but at one time were dangerous predators and were perhaps the reason for the evolution of the kea's cryptic coloring.

Schwing met his first kea in the wild while sleeping in a micro-tent, a one-person, two-and-a-half-pound affair. "The tent was ridiculously small—literally just big enough for me, with an inch or so between my face and the top of the tent," he recalls. For ventilation while he slept, he had left the zipper down a few inches. Before dawn, he woke with a start to a noise just above his head and found himself looking straight into the curious eyes of a kea, which had stuck its head through the zipper opening. "I'm not sure which of us was more freaked out."

It was the beginning of a love affair. Schwing was the first person to do a PhD thesis on wild kea. For two years he recorded their vocalizations and then grouped them into seven categories, ranging from that screeching piercing "hello" contact call, which carries well over long distances and in windy conditions, to a soft whistle, barely audible but with almost magical effects on the listener (Schwing dubs it the "cuddle call"). Some of the kea at Messerli use the whistle call with Schwing, but in the wild, it's extremely rare. Out of nearly twenty-five thousand calls he recorded, only twenty-four were this whistle. Schwing believes it's a very personal communication between one bird and another that seems to say, "I want to have close contact." A video he took of the whistle call shows a male perched on a bench. A female lands and moves toward him to preen him, but he kicks her away. She jumps back and then whistles quietly. He immediately lowers his head and allows her to come in and preen him. "We have birds in the aviary who can be quite rambunctious and bite hard," says Schwing, "but when another bird gives the whistle, they become very gentle." Schwing suspects these vocalizations are actually externalizations of the birds' internal emotional state.

While he was recording at one of his research sites, a lookout for sightseers, a kea stole Schwing's pen. When Schwing followed the bird to a bush, he found hidden beneath it a treasure trove of things pilfered from tourists—two pacifiers, four sets of eyeglasses, bits of antennas and rubber insulation from cars, and yes, his pen. One unfortunate Scottish tourist at the same site lost a wad of cash when a kea snatched it from his dashboard. What's cool about this, says Schwing, is that it happened after New Zealand had switched over to plastic money, which doesn't deteriorate. "You can't even rip it," he says. "So somewhere in the mountains, twelve hundred New Zealand dollars are lining the most gangster kea nest in the world."

These birds have a strong mischievous streak and are notorious vandals. At campgrounds in New Zealand's Southern Alps, kea have been known to dive beneath parked vehicles and steal their engine hoses or to gnaw holes in the tops of camper vans. During a field study, Schwing needed to get around, so he rented a car. "I knew enough to get full damage coverage," he told me, "but I don't even want to know what they thought when I brought it back. It was just ripped to pieces." He shows me a photo. "This is the antenna from that car. It took the kea three minutes to unscrew it. Not strip it off, not bite it off, *unscrew* it."

Like ravens and African grey parrots, kea have brains that are large relative to their body size and densely packed with neurons. "Nearly every aspect of their behavior shows intelligence," says Schwing. The aviary in Austria has a little hut with a touchscreen computer inside that allows researchers to investigate the cognitive capacities of the kea—for instance, testing their ability to complete tasks that require reasoning by analogy. If the birds push the right answer on the screen, they get a reward. "They all line up outside the little hut because they love to use the touchscreen," says Wein. For her master's project, Wein studied picture object recognition in kea, investigating whether the kea could transfer learning from photographs on the touchscreen to real objects, and vice versa. She taught the kea that one picture was positive and would yield a reward, and one was negative, and wouldn't. "They learned really quickly to transfer what they knew from pictures to real objects," says Wein, "like they scored 100 percent in just the second session."

Kea are also very good at learning from one another. One test of this involves a "tube removal" task. The inside of a small tube is smeared with a reward—butter, a big favorite for kea—and the tube is placed at the bottom of a long upright pole. To get at the treat, the kea must slide the tube up over the top end of the pole, which requires climbing up the pole while pushing the little tube with its bill. Kea accomplish this in just a few attempts, even if they've never seen the setup before. But when one kea watches another, it emulates the first bird's strategy and completes the task at first try.

Kermit is the wizard in the Messerli group. Give Kermit a problem and he will solve it within seconds, most often in a very rambunctious, kea-like style. The birds around him watch and imitate his actions. Not long ago, a team of scientists confronted Kermit and several other kea with a puzzle: a Plexiglas box with a reward on a little platform inside that could only be won by inserting a rod-shaped stick tool into an opening and maneuvering it toward the food to push it off the platform. For kea, with their curved beaks, this is no easy task. They can't hold a tool in alignment with their heads the way a New Caledonian crow can, so they have less control over the tip of the tool. Kermit nailed the solution with a unique routine that involved holding the tool in his beak laterally, then switching it to his foot and adjusting it with his beak until he could hold it in proper position with his foot, and then directing the tool through the opening with his beak and maneuvering it until he hit the reward.

"Kermit is very persistent and doesn't get easily frustrated," says Wein. "If he doesn't get it, then there's probably something wrong with your experiment."

These birds are smart, dogged, and very, very focused. In a small aviary set aside for studies, graduate student Thirsa van Wichen is testing whether kea like to work for their food. Studies show that most animals prefer some sort of foraging challenge. Cats, for instance, love food puzzles—gadgets that engage their hunting instincts and encourage physical activity—and this has been found to reduce their stress. The same is true for kea. They love the little puzzles that engage their foraging skills. Our presence doesn't faze the birds in the least; they are so totally focused on their task they don't even look up.

Once kea have learned how to solve a problem or use a gadget, they don't forget. Schwing brings out a wooden box he used in an elaborate experiment some years earlier to test the kea's cognitive skills and ability to cooperate. The wooden box is raised on legs and bottomed by a tray that drops down if properly triggered. In the experiment, the tray was loaded with treats—parrot pellets stuck there with cream cheese. Attached to the tray were four separate chains threaded through holes in the side of the box. To release the tray and access the treats, four kea had to pull the different chains simultaneously—and continue to pull together until the tray released. If only one bird pulled at a time, the chain would slide out without moving the tray. If any of the birds lets go too soon, the tray would pop back up out of reach.

The results were impressive. Kermit led the way, and after a little training, the other kea quickly learned to pull together—and to persist—so they could all get their treats.

A couple of years later, a film and photography crew from the National Geographic Society wanted to take photos of the kea cooperatively solving the wooden box and strings puzzle. "I didn't have high hopes," Schwing recalls. The kea hadn't seen the puzzle box in over a year. But he took what he thought were his four best birds and put them in a waiting compartment while the crews set up the cameras.

"It was breeding season, so there was some tension—they were chasing each other around in there," he says. "And I thought, 'Oh, this is not looking good.' But the moment I opened the gate for them to get access to the box, they fell in line, one after another, not even a body length in distance between beak and tail. And in this group of four they walked right over to the box, past the photographers, past the lights, past everything, walked right up and found their places and pulled on the chains.

"We did this several times so the camera crew could get all their shots. The photographer ended up lying down with his head so close to one kea that he was blocking access to the chain. The bird walked over and just pushed the photographer's head aside with his foot so he could get to the chain and pull on it.

"They had no trouble remembering what they were supposed to do," says Schwing. "And they were just so insanely focused."

Now, when Schwing sets the box on the floor of the aviary, all of the kea come running willy-nilly and gather around the box in chaotic confusion, the young, inexperienced birds grabbing the chains two at a time, making a mess of it while the adults let them play. It's parrot pandemonium.

In New Zealand's Fiordland National Park, the kea have figured out how to spring the traps set for stoats, an invasive mammal plaguing the island. The traps are housed in solid wooden boxes designed to deter such damage, but the kea have figured out how to spring the traps with just the right size stick or by undoing the lid screws. Sometimes their efforts are rewarded with eggs or other bait; sometimes just with a loud bang. They've even dug up the traps and rolled them over bluffs.

Such boldness and curiosity can get a bird into trouble. "Sometimes they're too smart for their own good," says Schwing. "A few months ago, a man shot at seven kea on his property because they had 'uninstalled' the satellite dish from his roof."

Like ravens, kea are opportunistic, "extractive" foragers—another trait common to birds that play. However, until lately, their foraging behavior in the wild has been largely a mystery. In the 1950s and 1960s, a park ranger, J. R. Jackson, conducted the first field studies of kea, collecting copious data and observations on their territory, food, and population. "He was extremely devoted," says Schwing, "to the point where he would hang out at the garbage dumps near Arthur's Pass where kea congregate, creating a kind of blind by climbing into a burlap sack and looking through the edge. He wouldn't move, even when people came up and started kicking the sack." Jackson discovered that the birds nest on the ground in tunnels and cavities beneath logs, rocks, and ledges, and there's often a well-worn path leading to the nest entrance. There are five or six papers to Jackson's name, but the story goes that he had stacks of notes for papers he never got a chance to publish before he went missing in the mountains.

Alan Bond and Judy Diamond of the University of Nebraska studied the kea at Arthur's Pass beginning in the 1980s and wrote a fascinating book about the "bird of paradox." They were the first to note the kea's unusually high rates of object and social play compared with other species and explored their breeding and feeding habits.

Kea exploit a vast cornucopia of food sources, possibly the most varied of any bird. They eat lots of different things because they have to—alpine environments offer sparse food resources—so the birds have become "truly omnivorous," write Bond and Diamond, "with a breadth of interest in plant and animal foods that roughly matches that of humans." Born botanists, they eat an enormous variety of plant species and—as we know—have figured out how to dissect poisonous plants to get at just the edible parts. They're carrion feeders, too, and will also eat live flesh when they can, including from the rumps of sheep.

"I'd like to say they don't do this," says Schwing, "but they do." This has earned the birds the enmity of sheep farmers. Rumor has it that kea have a taste for kidneys, and that's why they chew away at the rump. But Schwing doesn't buy this. "The birds just aim for an area that's unprotected," he says. "To this day, the sheep in New Zealand are trimmed of their long tails. The tails, tied up when they're lambs, just fall off at some point." (Which makes for some strange and morbid scenes, he says, fields of sheep tails just lying around. "The first time you see it, it's like, 'Wow, those are the largest caterpillars I've ever seen!' Then you realize . . . 'Oh wait a minute . . . '")

"The problem is, the long tail of a sheep has a function, which is to chase away things from your back that you can't reach," says Schwing. "I think the kea are zeroing in on the one spot where the sheep can't get at them and where they can actually get into the soft, tender flesh between the ribs and the hips. That's where they hold on and chew."

From 1890 to 1971, the government offered a bounty for dead kea. Methods for killing the birds ranged from poisoned bait to a special kea gun, according to George Marriner, who wrote a book on the bird in 1908. Marriner described kea as "the most inquisitive bird alive" and "a most lively and interesting companion," and in the same breath, boasted of killing them "at all hours, from the first streak of dawn to the last faint glimmer of daylight." Before they were finally granted full protection under New Zealand's Wildlife Act in 1986, more than one hundred fifty thousand kea were slaughtered.

"Cluing into the kea's smarts at social learning might actually benefit the sheep farmer," says Schwing. The best way to deal with kea attacking your sheep? "Remove the young male that has started the practice of feeding on the sheep before it spreads as knowledge within the group," he advises. "If you move that young male—and it's almost always a young male, what we call an 'innovator personality'—sheep and birds can live again peacefully until the next young male comes along to innovate."

If intelligence and innovative foraging are key ingredients in animals that play, so is an extended juvenile period. Kea tick off this box, too, with a very long "childhood" of four to eight years. "That's incredibly long, even for a parrot," says Schwing. Kea are not as long-lived as most parrots. An Amazonian parrot will often live eighty or ninety years. Sulphur-crested cockatoos have an average life span of sixty-five years but can live to one hundred twenty in captivity if they're well cared for. "Kea live only forty years max," says Schwing. "So even if their juvenile phase is just four years, that's 10 percent of their maximum life expectancy—that's a huge amount of time compared to most other creatures."

As if on cue, a juvenile named Skipper, born the year before, sidles up to us with bandy-legged bravado, brash and adorable with his feathers all puffed out. You can tell he's a young bird because he has yellow around his beak, a yellow ring around his eyes, and a little yellow crown. He swaggers and struts as if he owns the place.

Which he does.

This is another cool thing about these birds, Schwing says. During their long childhood, juveniles have a favored position in kea social hierarchy. You see it even with the fledglings. When lunch is served—a mixture of seeds, fruits, and vegetables laid out on little platforms stationed around the aviary—the adults settle into peaceful eating. A fledgling arrives, pokes around on the ground a bit, and then, spotting better fare above, flaps up to the platform and takes the food right out from under the adults' beaks, literally while they're eating it, with complete impunity. I watch this happen with different adults, different young birds, three, four, five times. Not once does an adult fight off the juvenile taking its food. The young birds always get their way.

Schwing and Wein sometimes have to hand-raise baby kea and

introduce the young birds to the group. "In other parrot species, if you hand-raise chicks and put them in the group, they will more likely than not be killed because there's always this in-group/out-of-group differentiation," says Schwing. "You have to slowly habituate them to the group over months—putting them in a cage in the group—and gradually introduce them this way." But with kea, it's the opposite. "The moment we give the adult kea access to a young bird, the males almost fight with one another to see who gets to feed these unknown, unrelated birds. The first day the fledglings arrive in the group, they're overwhelmed with food. They start beating their wings to get rid of the adults because they just can't swallow any more food."

Contrast this with the strict pecking order in raven society illuminated by a story from a knowledgeable birdwatcher, who was sitting on the deck of a friend's house in the woods in California. The house has a ravens' nest in a large tree behind it, and the friend had been training the resident ravens to come to him and take dead mice from a stick. The friend writes:

> One day, a smaller raven flew in and took a mouse into its crop. As soon as the raven swallowed the mouse, a larger raven, perhaps the nesting female, came down screaming mad at the young raven, who she made disgorge the mouse. She picked it up and took it up to the nest, leaving the smaller raven sitting on the edge of the deck with his head down. She came back down, still screaming, and hit the smaller raven multiple times with her wings along both sides of the head. Another incident of feeding hierarchy in raven groups—the younger raven was clearly supposed to have shared the dead mouse with the nestlings instead of taking it himself— and paid the price.

"Kea just have this built-in 'juveniles go first' mentality," says Schwing. "We see it in captive populations; we see it in the wild. Juveniles and fledglings can walk up to any bird and just take food away from them."

Once a young kea hits puberty, however, everything changes. When a bird approaches sexual maturity and loses its little yellow crown and eye ring and its feathers darken, the special treatment ends, and down it drops to the bottommost rung of the social ladder. "Kind of takes the wind out of their sails for a while," says Wein. But the exceptionally long juvenile period and the privileged position of young birds in the group structure offer them plenty of spare time to indulge in almost constant play.

It's not just the juveniles that frolic nonstop. The adult birds are equally plucky and playful. They seem to have taken to heart the adage sometimes attributed to George Bernard Shaw, "We don't stop playing because we grow old; we grow old because we stop playing." Everywhere in the aviary, kea are deep in rough-and-tumble horseplay. They're locking bills and twisting, rolling around and wrestling, one kea lying on its back and the other jumping on its stomach like a kitten. They're sneaking up and grabbing each other by the beak or feet, jabbing at each other with what Schwing describes as "kung fu" kicks. Frowin, a mature adult, is playing with a leaf, tossing it in the air and catching it. Kea may love new things, but they will also pick up something that's been lying around forever and just start throwing it around.

In the mountains of New Zealand, Schwing saw plenty of this so-called ground play. He expected the tussling and tossing of objects. "It was pretty much the accepted view of what constitutes kea play," he says. "That's because Bond and Diamond chose as their study site a garbage dump in the low country where kea liked to congregate. They come down to hang out where people and food are." The researchers saw kea pulling bones or pieces of cloth from both ends, tossing rocks, bottle caps, and walnuts in the air, and a group of fledglings that apparently found an hour of play in a long piece of surgical gauze, "walking around with it and periodically hopping, jumping, and foot-pushing."

But as Schwing points out, garbage dumps are usually located in valleys where there is minimal air exchange with surrounding areas so the smell won't spread. That means there's little to no wind.

When Schwing was recording and observing the wild birds in their natural habitat in the high mountains, he saw that the most popular form of play among wild kea was aerial—chasing, foot-kicking, flying acrobatics. His study site was Deaths Corner lookout in Arthur's Pass National Park, a popular spot for kea. The lookout earned its name from a huge crevasse that drops off two hundred meters and allows a view extending almost to the ocean. Next to the parking lot is a huge electrical tower, where the kea like to perch. It's a good foraging location and also a popular spot with tourists, who tend to offer snacks. But most important, says Schwing, "the big cliffs and huge valleys create a lot of updrafts, which give kea the ability to fly really fast and really far with minimal energy expenditure." One bird will land on the back of another while airborne, or they'll chase each other in wild, careering flight, or perform loop-the-loops, sometimes momentarily flying upside down, or do spirals or wheelies side by side, then land right where they started.

They also hover, riding the updrafts but staying stationary, which involves tilting their wings ever so slightly second by second. "The game is to stay in the same place," says Schwing. "You could argue that this is not play but hovering to survey the landscape from above, like a hawk does. Except that they do this just five feet off the ground, so they're expending tremendous energy to stay in a stable horizontal position and gaining nothing in terms of a view. You don't realize how much power it takes to do this until another bird whips by in the wind, and the hovering bird decides to join it. A tiny shift in the position of its wings and pop! It's gone."

What really enthralled Schwing was the special call these birds utter only during play, a sweet, funny, squealy warble. Visual play signals are common in animals—play faces of primates, play bowing in dogs and Australian babblers. But play sounds are rarer, observed only in rats, squirrel monkeys, chimps, and cotton-top tamarins. To explore the kea calls in the wild as part of his PhD studies, Schwing set up speakers and video recorders at Deaths Corner lookout one cold and dreary winter. He deliberately chose the worst season for the study and went out in the worst weather, gloomy, cold, and drizzly, when no bird in its right mind would be in the mood to frolic. Through the speakers, he played five recorded sounds for five minutes apiece, the warbling play call; two other kea calls; a call of a common bird in the area, the South Island robin; and a simple standardized tone.

The little videos of the results are a hoot. Two adult kea just hanging around on a stone wall, minding their own business in the misty drizzle. The recording of the warbling play call is heard in the background, and suddenly the birds look at each other, burst into squeals, and launch into play, becoming exceedingly silly, chasing each other, flapping up and down, picking up rocks and flinging them, playing hard for the full five minutes of the recording.

"With the warble calls, play just went up drastically," says Schwing, "500 percent." Kea of all ages jumped in, with some of the most vigorous and longest bouts in mature adults, including between males and females. Play between fully mature animals of the opposite sex is rare in animals—and if it occurs, it's usually part of courtship behavior or to strengthen social bonds before a hunt. But in kea, it's not about food or sex. It's about play.

The kea didn't look around for the source of the sound, says Schwing, which suggests to him that the call isn't so much an invitation, "Come play with me!" as it is an inspiration: "Playtime!" The instant the play call recording stopped, so did the play.

"Hearing the play call boosted the birds' intrinsic motivation to play," explains Schwing. It's the same kind of emotional contagion that affected the raven nestlings, and it's a lot like the infectious giggling of close friends. Schwing is quick to say he's not actually calling the warble laughter. But he sees it as a close analogue because both play call and human laughter, especially in small children, have similar effects—they're positive behaviors that trigger positive behaviors in others. Hearing people laugh puts us in a better mood and gooses the humor we perceive in things. Play the kea play call to a group of people who have never heard a kea in their life, and they'll laugh and say, "Oh, that sounds happy!" Play it to kea, and they'll throw things, chase each other, or if no other birds are around, start playing by themselves.

Why would kea have a play call that is so contagious? What use is it to a parrot to rope others into mirth?

Schwing is focused on working this out—and on understanding why kea are so playful to begin with. He thinks it has to do with their social system and its almost utopian nature. Kea don't have the normal hierarchical behavior you see in other highly social animals. In most social bird communities, status in the group is established by fighting. Kea don't have the same sort of pecking order, says Schwing. They don't fight for dominance. "In four years of fieldwork in New Zealand, I never saw two adult kea fight," he says. "Not once." In fact, he saw only one incident that was anything like a fight, and it was between two juveniles at a carcass who both wanted to eat from the same opening.

In fact, kea may not have a true hierarchy in the wild, but rather a more fluid social organization. "If you go to what I call a hot spot in the mountains, where you have conglomerations of thirty kea, they mingle in very fluid groups," he says. "You might have two birds arrive, then six birds, then three birds fly away, five birds come in, four birds fly away. They're really mixing it up." In other highly social species, this sort of mingling would create tension and fighting.

You have this kind of exchange in other species, says Schwing, "but not in such abundance and at such an incredibly fast pace. With ravens, new birds may come in and others leave—but over a period of weeks. With the kea, we're talking hourly changeovers of maybe half the group. If the kea had a true hierarchy, they would be fighting all the time."

Instead, Schwing argues, they play. They use cavorting, clowning, wrestling, and horsing around as substitutes for hierarchical fighting. By playing together so extensively, they create a tolerance in their groups that obviates the need for a hierarchy based on fighting. In other words, they use play as a social facilitator. In this way, kea play is a lot like teasing in humans, which is often a way of testing limits or tolerance. As Amelia Wein points out, kea seem to have a nose for doing the very thing that's most annoying in a given moment—pulling the threads from the mop students use to clean their aviaries, plucking holes in your shirt, untying your shoelaces. In their capacity to tease, kea may be revealing a key bit of social acumen. Teasing and clowning in humans and other primates depends on an awareness of other minds. "You can't tease other people unless you can correctly guess what is 'in their minds,'" writes the psychologist Daniel Stern. It's fun because you realize how your target feels. In other words, social play requires keen social intelligence.

It also contributes to it. As play researcher Sergio Pellis has shown, play changes the social brain. Playing with others modifies the connections between neurons in the prefrontal cortex, and these changes likely mediate the development of social competence in any animal. Birds deprived of play in youth will not play in adulthood—and may struggle to fit into social groupings. One kea at Messerli occupies her own little aviary. She was raised in isolation in a zoo and never experienced play. Now she is unable to play and also has trouble being with other kea. A single bird does not a theory make. There's no evidence of cause and effect between this bird's play deprivation in her early years and her current social unease, and there are lots of other possible reasons to explain it. But the example is intriguing.

If kea play behavior is a kind of social facilitator that allows for tolerance and actually increases it in a group, muses Schwing, then having the warble play call, "a sound that travels and has a direct effect on the emotion and tolerance and playfulness of all the birds that can hear it . . . would exponentially increase the social facilitation effect."

At some level, then, kea seem to understand the old clown saying, "Laugh and the world laughs with you . . ."

Halfway around the globe, at the Willowbank Wildlife Reserve in Christchurch, New Zealand, Alex Taylor and Ximena Nelson are working to understand whether the warble call is truly analogous to human laughter and whether there might be deeper positive emotions underlying the call.

Willowbank is a wildlife reserve that focuses on maintaining populations of endangered native species for educational purposes and for breed-and-release programs. The thirteen kea there have become subjects in a series of studies designed by Taylor and Nelson to explore the play warble. Does it function like human laughter? Does it have similar effects on the way a kea feels and on its well-being? Does it defuse conflict and reduce stress?

Nelson is a native Kiwi who has been investigating kea since 2010. Taylor is a relative newcomer to the species. He has worked for years with the New Caledonian crow, considered by many to be the most intelligent bird in the world. The crows are supersmart and champion tool users, says Taylor, but they have a very low tolerance for novelty, and they don't play at all. "You couldn't get two birds more different when it comes to playfulness. Kea are absolutely at the other end of the spectrum." But Taylor views the warble call and play behavior of the kea in the same way he views tool use in the New Caledonian crow, as a means to explore a bird's mind.

Taylor's first impression of the kea was pretty much the same as everyone's, he says. "Suddenly you have this very intense interaction with an animal that's trying to climb on your shoulder or steal your shoelaces or nibble your ear or pull a strap off your bag or get your camera or pencil. I can't think of another animal that has this huge love of novel things and interacting with them. Maybe a monkey or great ape? But certainly, no other bird."

Willowbank is a magical place, says Taylor. "I send students down there to see how good they are with the birds. Within five minutes, they're dealing with not just one kea, but five, six, seven, eight birds, wanting to interact with them. What better job interview for a young student?" The favorite research subject there is a kea named Kati, who is missing her top beak. Schwing and Wein rescued her from the mountains after she had gotten her beak caught in a trap and lost the upper mandible. "She would have had no chance in the wild," says Schwing, "so we called in to Willowbank and said, 'We have this perfectly healthy little kea, should we catch her?' They said, 'Absolutely, Raoul, go ahead and catch her.'" Later it was discovered that Kati was a male and renamed Bruce, but everyone still calls her Kati. Now she's the star at Willowbank, a perfect ambassador for kea, and an eager participant in scientific studies. "She's an amazing kea to work with," says Taylor. "She's really interested in working on the experiments, really bold."

Taylor sees Kati's quirky, charismatic species as a fascinating evolutionary riddle. "How similar are these birds to humans when it comes to their emotions?" he asks. "Kea are extreme birds, but if we can understand how they're thinking and feeling, it will open up the question for other species."

Scientists suspect that laughter may have evolved from the breathy panting and labored breathing during play fighting and tickling between our primate ancestors and was used as a signal that everything is fine, now is a good time to socialize, play, explore together. It dates back at least to the common ancestor of humans and apes, some ten million to sixteen million years ago, but it probably goes back a lot further than that. Neural circuits for laughter reside in parts of the brain that are very old indeed, dating back to our earliest mammalian ancestors.

When neuroscientist Jaak Panksepp watched rats play at his lab at Washington State University in the 1990s, he noticed that their rambunctious antics were accompanied by bizarre chirps and squeals, known as ultrasonic vocalizations. He wondered, could these sounds emitted during play be an ancestral form of laughter? He and his colleagues explored the brain circuitry involved in the rat vocalizations and found that it corresponded with a reward pathway in an ancient part of the mammalian brain known to be activated during feelings of optimism, joy, and enthusiasm. Birds have the same pathways. As Panksepp writes, "Ancestral forms of play and laughter existed in other animals eons before we humans came along with our hahahas and verbal repartee."

Laughter in humans is far more than just a reaction to something funny. It's a basic communication tool, a building block of society, defusing social tension and building social bonds. It also enhances both mood and well-being. "When we laugh, we feel happy and optimistic," says Taylor. "We act friendlier, don't argue as much, feel less stressed, and even become healthier." In a series of ambitious experiments, he and Nelson are looking for these telltale "signatures" of human laughter in the kea warble. Do kea feel happy when they warble—or at least more optimistic? Does the call enhance social bonds and ease conflict? Does it reduce stress and offer other health benefits? And is it affected by environmental conditions? These are all indicators of a meaningful analogy between the warble call and human laughter.

For starters, the pair plans to test kea to find out whether warbling might shift the birds' mental state to make them more optimistic, as laughter does for us.

This may seem like anthropomorphizing, projecting human experience onto an animal's behavior, but it turns out there is a solid, well-established test for optimism in animals. Here's the kea version: The bird is taught that a big box always contains food and a little one does not. Then a midsize box shows up. "An optimistic animal will take a 'glass half-full' perspective," says Taylor, "and assume that the medium box is like the big one and will run right up to it." What happens in this situation when a kea hears a warble call? Does it approach the intermediate box more quickly?

Taylor and Nelson are also devising experiments to create mild conflicts between kea to see if playing a recording of the warble call defuses tension between birds, and investigating whether kea warbling and mood is affected by the environment.

We know that some people get seasonal affective disorder when the daylight hours are short. We know that sunlight tends to boost human mood. Nelson and Taylor wondered whether a kea's positive emotions and the urge to play might likewise be affected by the natural environment. Nelson is a keen skier and spends a lot of time outside. Over the course of her time studying kea in the field, she noticed that on a sunny day, after a fresh snowfall, the birds do what we might do—they warbled and played more often than on a gray, rainy day.

So here's our question, says Taylor: "Do we see kea warbling and playing more in nice environments and conditions?" The team is seeking answers in the field—they've sent a PhD student into the mountains to observe and record how much the kea warble and play when it's bright and sunny versus when it's dreary—and also in the lab, treating kea with bright lights in winter to see if this triggers more warble calls, play, and optimism.

This ties in very nicely with kea welfare and their sense of well-being, says Taylor. He and Nelson plan to take their research a step further to ask if the kea warble has the same physiological effect in the birds that laughter does in humans, lowering levels of the stress hormone cortisol and, perhaps, improving healing—research with some practical applications.

Like children, kea suffer from lead poisoning. They're drawn to the lead in old housing and trekking huts, Taylor says, for the same reason children are drawn to it in paint chips, because it tastes sweet. But for birds, as for babies, lead is highly toxic, shutting down a bird's digestive system and its immune system. The problem is widespread, and at rehabilitation centers in New Zealand, dozens of lead-poisoned kea are treated with antibiotics and anti-inflammatories. Taylor and Nelson wonder if listening to daily warble calls might accelerate their recovery.

"It would be amazing to find there's an overlap between human laughter and the kea warble call in terms of its function, when it's expressed, how it makes the kea feel," says Taylor. "If we can show this— that the kea warble call makes the birds feel more optimistic and friendlier, that it's affected by the weather, that it actually decreases their stress and boosts their health—that would raise really interesting questions about the evolution of emotion, and, we hope, spark more focus on bird emotion in our field generally. It might also shift the public's perspective on the kea from destructive pest to emotionally intelligent parrot, capable of joy and laughter, which would help with conservation."

With only an estimated three thousand to seven thousand individuals left in the wild, the kea is classified as endangered. One sideline to Nelson and Taylor's research on kea play promises to address at least one threat to the bird.

Not long ago, the staff members of the New Zealand Transport Agency were baffled to find traffic cones used during road construction scattered randomly around the entrance to the Homer Tunnel, far from where the road crews had carefully placed them. Footage from video cameras captured images of several young male kea pushing the cones around and repositioning them, sometimes leaving the cones in the middle of the highway just outside the tunnel.

The birds might just be having a little fun, says Taylor, but there also may be method to their madness. When the cars slow to avoid the cones, the kea come to the windows to investigate and poke around for food and toys. This kind of interaction between kea and cars occurs in other places on the South Island, too, wherever there are birds and stopped traffic. "The problem is that kea wandering into the road get injured or killed," says Taylor, "and their interaction with people is not what you want."

In 2018, Taylor and Nelson went to the Homer Tunnel just outside Milford Sound to see the situation for themselves. "My heart was in my throat," recalls Taylor. "You've got one line of traffic stopped, all of these people queueing up to go through the tunnel, and the kea are in the middle of the road, just hanging around, looking for food. They're on the roofs of cars, trying to squeeze under them, trying to get in the windows. There's traffic coming out of the tunnel, massive 18-wheelers thundering through, and the kea step out of the way, but barely."

Taylor also watched an unnerving exchange between a woman and one highly persistent bird. "The kea is on her roof, and her window is down but she can't get her camera pointed up there," recalls Taylor. "So she gets out of the car to take a picture, and the kea drops down from the roof and wanders over to the open door. And he really wants to get in the car because he knows there's food in there. Without thinking, the lady slams the door, and it misses the kea's head by a couple of centimeters. The bird barely flinches and then goes for the open window. And then she kind of panics and bats at it. "The whole situation was a disaster," says Taylor. "Exactly the kind of negative interaction you don't want."

"This kind of thing happens all the time," says Nelson. "It's not the fault of the kea, but the result of decidedly unhelpful human behavior."

In the past, New Zealand's Department of Conservation has dealt with inquisitive kea causing trouble at forestry projects by providing the birds with a ladder to distract them and keep them away from dangerous and expensive equipment. This gave Nelson an idea: build a kea jungle gym by the roadside and make it more appealing and alluring to the kea than the cars, by putting in a variety of toys—swings, climbing frames, ladders, puzzles, and flotation devices—and switching them up a lot, rearranging and refreshing them regularly to hold the birds' interest. "The idea is that the kea will be drawn there and so will the tourists who want to watch them and get pictures," says Taylor. It helps that kea are motivated more by play than food. "If there's any species in which play and new objects are more important than getting a crisp, it's kea."

"The challenge here is to figure out what is essentially catnip for kea in terms of objects they absolutely adore playing with," says Taylor, "and then somehow to find a constant stream of novel variants of these toys." The plan is to try to turn this into a citizen science venture, the Great New Zealand Kea Jungle Gym Project, Taylor quips, "and get school kids and anyone else to send us designs for toys they think will be really engaging for kea. And then we could take those designs and build them up and see how the kea interact with them."

What a fun bit of interspecific reciprocity: people playing around with objects to find suitable toys for kea to play around with. And kea, in turn, providing entertainment for people at play.

Love

Chapter Nine

SEX

I've always loved the refined look of male mallards, their shiny iridescent green heads ringed by a narrow white band, and those crisp violet speculum patches fringed in white on the wing. But unmated
males among these dabbling ducks have a well-deserved reputation for
truly dreadful mating behavior. Groups of drakes may force themselves
on an unwilling female, sometimes attacking with such violence and in
such numbers that they kill her in the process. Wood ducks are also
known for this brutal practice, dozens of males try to mate with a single
female.

Contrast this with the billing of Atlantic puffins, a gentle rubbing together of beaks in anticipation of sex. Or the sweet-seeming presentations of flower petals by male superb fairy-wrens. Or the tender mutual
attentions of Fischer's lovebirds, those vibrant pint-size parrots native to
Tanzania that gave us our expression for openly affectionate couples.
Lovebird pairs snuggle, gently preen each other, and nibble beaks, the
bird equivalent of kissing. After separation or stress, they feed each other
to affirm their bond. When a male approaches his mate, he sidles back
and forth energetically, bobbing his head up and down, twittering, and
then regurgitates some food into her mouth. If a mate dies or disappears,
the other partner will pine for it. This is true, too, for mated pairs of those
maligned mallards. When they form pair-bonds, they stay together for

the season and solidify their bonds with precopulatory displays of head bobbing, nodding, and shaking.

For sheer variety, the sex lives of different bird species are hard to beat.

At first blush, their reproductive organs seem all alike. Both male and female have a cloaca, an opening that in the male swells during the mating season, projecting outside their bodies. When birds mate, they briefly rub together their swollen cloacae, allowing the male's sperm to move from his cloaca to hers and then travel up her reproductive tract to fertilize her egg. (Bird cloacae also have a decidedly less sexy role: to excrete urine and feces.)

Most birds have no penis. But there are exceptions. Several species of ducks, geese, and swans have an organ like a human penis, which is inserted into the female. These waterfowl belong to the 3 percent of living bird species that retain the phallus found in their reptilian ancestors. Some ducks—such as those brutish male mallards—have impressive counterclockwise corkscrew-shaped, snakelike phalluses that grow as long as their bodies, which they use to deposit sperm as far as possible inside the female reproductive tract, to better their chances of fertilizing her eggs. The length correlates with the degree of forced copulation that males impose on female ducks, says biologist Patricia Brennan.

It's the evolutionary upshot of a sexual arms race. Until recently, all the focus had been on the duck penis, but really—as Brennan discovered—this anatomical story is a story of female agency, even in the face of extreme sexual violence. Brennan has found that females have evolved their own rococo genitals, a spiral-like reproductive tract that winds clockwise, in the opposite direction from the male's penis, and includes up to three branches with blind pockets that make it harder for sperm to reach her eggs. "Female mallards have a say in which sperm fertilize their eggs," she says. "As many as 35 percent of all copulations with a mallard female are forced by unwanted males, yet these males sire only 3 percent to 5 percent of her offspring." With a forced copulation, the female keeps her genital tract tight, blocking the male's lengthy phallus or forcing it to divert down one of those dead-end alleys, so his sperm can't fertilize her eggs. If she actually wants to mate with her partner, she can relax the walls of the tract and allow his semen passage. She can also eject sperm from her cloaca by defecating right after sex.

As for the act itself, many birds copulate quickly for only a second or two, a "cloacal kiss" that effectively transfers sperm. Notable exceptions include the red-billed buffalo-weaver and the aquatic warbler, which may couple for more than a half hour. The male warbler "clings to the female's back so that the pair hop around together in the vegetation like a couple of mice," writes British ornithologist Tim Birkhead. The vasa parrot, a native of Madagascar and the nearby Comoros islands, may hold some sort of record. One pair was observed copulating for a full 104 minutes, cloacae locked together, the male's tail curled up under the female's tail, her wing extended over his body.

Some birds copulate just once, achieving fertilization in one quick go, while goshawks couple as many as six hundred times for a single clutch of eggs. Why so many tries? If you're a bird of prey and have to spend long periods hunting, away from your mate, it's harder to guard her from having sex with other males. Multiple matings may dilute any semen acquired from other males in your absence.

Most birds do the deed out in the open, on a nest or patch of ground or branch or, occasionally, in the case of oxpeckers, on the back of another animal, like an African buffalo. Common swifts, birds that spend nearly their entire lives aloft, eating and sleeping on the wing, mate mostly at the nest, but have been known on occasion to copulate in flight, male just above and behind the female. Gulls do it on the beach.

By contrast, Arabian babblers often go to great lengths to hide the act of sex, copulating out of sight of their group members, just as humans do.

Concealing sex is thought to be pretty consistent across human cultures. The Yanomami of Venezuela, for instance, arrange trysts away from villages and out of the public eye. Married Malekula couples of

Melanesia return home from a rendezvous in different directions. And Mehinaku couples in Brazil have sex only in secret locations. In fact, the penchant for private sex has been considered a universal human trait and one that may have influenced the evolution of human emotions and advanced cognition.

Now along comes research by ethologist Yitzchak Ben Mocha suggesting that Arabian babblers also take pains to conceal their intimate moments. Other animals occasionally hide their sex acts—chimps and baboons, for instance, as well as a few bird species, dunnocks and alpine accentors. But it's usually subordinate males that do so to avoid being harassed by more-dominant alpha males in their groups. In Arabian babblers even the most dominant group members conceal their copulations from the rest of the social group. As far we know, says Ben Mocha, this is the only species other than humans where individuals conceal sexual interactions that are not expected to be interfered with by other members of a social group.

These birds, which live in the harsh deserts of the Arabian Peninsula, breed in long-term cohesive cooperative groups with a dominant pair raising young with helpers that pitch in to feed chicks, defend territory, and ward off potential predators. Ben Mocha found that the dominant pair not only regularly hides its sex from the view of the rest of the birds in the group but also uses some sophisticated cognitive skills to do so. A dominant bird will slip off and head for a spot where it's visible only to its partner. The bird signals the moment for the meeting by plucking up in its beak some arbitrary object nearby—a stick or eggshell—and waving it around at its intended, then nodding its head slightly, taking scrupulous care to signal only when no other babbler is looking its way. In one instance, a female was perched on top of a bush and the rest of the group was foraging to one side of it. A male moved to the opposite side of the bush and signaled at the female from there. This suggests that babblers may be able to distinguish their own visual perspective from an alternative point of view—an advanced cognitive skill known as perspective taking.

According to Ben Mocha, the behavior may also entail tactical decep-tion, using a signal or display to mislead or deceive another individual. The object that a male or female babbler picks to use as a signal is most often a small twig or leaf that could well be used for some other purpose, like nest building—a way to conceal from the others that it's actually sig-naling. When another babbler approaches, the signaling bird drops the object and acts as if there's nothing interesting going on. "The observing bird cannot know then if it was an invitation for mating or, for example, searching for nesting materials," explains Ben Mocha.

Once a babbler receives the signal and decides to cooperate, the two birds sneak away together—when no one else is watching—to copulate behind a bush or tree. Then they'll hop back onto the scene seconds later "in a very normal way," Ben Mocha explains.

"I don't think they try to hide the fact that they've had sex—just a view of the act itself." This is another way the babblers are like humans. "Think about a human wedding," says Ben Mocha. "The bride and the groom invite many guests to the ceremony. Everyone knows what will happen after the ceremony, but no one actually sees it."

Why would these birds go to such trouble to hide their sex acts?

"We don't know," says Ben Mocha, "in either birds or humans. We're still not sure why humans conceal socially legitimate sex (that is, sex that does not violate social norms and therefore is not expected to be inter-fered with by other humans). Opening the door to cross-species compar-ison may shed light on the evolution of this behavior," says Ben Mocha. "I believe that humans conceal legitimate sex for reasons similar to the babbler," but that remains to be studied. "Our evidence shows that bab-blers don't conceal sex to avoid predation or interference by subordinates or to signal their dominance," he says. They may do it to avoid "teasing" their helpers, whom they rely on to raise their offspring—a way of pre-venting social tension and maintaining unity in the babblers' social group, ensuring the cooperation of the helpers who don't get their own chance to mate.

The babbler's concealing of sex, then, may be another example—like play in kea or facial signaling in red-billed queleas—of the skillful ability of birds to balance and negotiate social relationships to keep the peace and maintain cooperation. Perhaps not far off from its purpose in humans.

Our view of bird sex has changed radically over the years. Behaviors we used to think of as rare or deviant we now understand are common.

Adélie penguins deserve a special mention in this regard. The penguin's sexual habits were first reported by George Murray Levick. A staff surgeon on a British expedition to Antarctica in 1910, Levick spent twelve weeks observing four big colonies of Adélie penguins living on Victoria Land. He published several scientific accounts of the penguins, as well as a little illustrated book called *Antarctic Penguins*, which offers a charming, if antiquated, introduction to these inquisitive birds: "The Adélie penguin gives you the impression of a very smart little man in an evening dress suit, so absolutely immaculate is he, with his shimmering white front and black back and shoulders. His carriage is confident as he approaches you over the snow, curiosity in his every movement. When within a yard or two of you, as you stand silently watching him, he halts, poking his head forward with little jerky movements, first to one side, then to the other, using his right and left eye alternately during his inspection."

Levick noted that males claim a territory, gather rocks to build a nest, and when the females arrive, are thrown into a kind of frenzy. They toss back their heads, beak to the sky, and discharge a string of hoarse trills and squawks, "a volley of guttural sounds straight at the unresponding heavens," he writes, "a *Chant de Satisfaction*." Then come a male's overtures to a hen. "He would, as a rule, pick up a stone and lay it in front of her . . . rise to his feet, and in the prettiest manner edge up to her, gracefully arch his neck, and with soft guttural sounds pacify her and make love to her. Both perhaps would then assume the 'ecstatic' attitude, rocking their necks from side to side as they faced one another, and after this a perfect understanding would seem to grow up between them, and the solemn compact was made. It is difficult to convey in words the daintiness of this pretty little scene."

Not so dainty were what Levick considered the penguins' "depraved" sexual behaviors. These shocked Levick so much that he wrote up his observations in Greek so that the vast majority of people couldn't read what he considered the offending passages. A sign of the times: Even thus redacted, the four-page report on this aspect of their behavior was deemed unfit for publication with the official expedition reports in 1915. It languished in the Natural History Museum at Tring, in Great Britain, until 2012, when Douglas Russell and his colleagues discovered it and found Levick's observations accurate, valid, and deserving of publication, despite his outdated views and occasional abandonment of scientific objectivity.

Levick was particularly appalled by the behavior of unattached males, "hooligan cocks," he called them, "whose passions seemed to have passed beyond their control." He wrote with a kind of gasping indignation of males mating with each other, with injured females, with chicks, with dead penguins, even with the ground itself:

I saw another act of astonishing depravity today. A hen which had been in some way badly injured in the hindquarters was crawling painfully along on her belly. I was just wondering whether I ought to kill her or not, when a cock noticed her in passing and went up to her. After a short inspection, he deliberately raped her, she being quite unable to resist him.

This afternoon I saw a most extraordinary site [sic]. A Penguin was actually engaged in sodomy upon the body of a dead white throated bird of its own species. . . . There seems to be no crime too low for these Penguins.

Levick had no context for the penguins' behavior. The truth is, plenty of birds interact sexually with dead members of their own species. The behavior has been noted in bridled terns, European swallows, sand martins, and Stark's larks. It's not at all the same as a human having sex with a corpse and can be explained in the context of our modern understanding of animal behavior.

Well, sort of.

There's even a term for it, *Davian behavior*, coined by Robert Dickerman—known as the biologist who gave necrophilia a good name. Dickerman speculated that the "lordosis" posture of a dead ground squirrel, with back arched downward, might release the copulatory drive in a sexually aroused male. And indeed, when biologist David Ainley conducted an experiment on Adélies with a dead female penguin frozen into position, he noted that older males found the female corpse irresistible.

However, a recent study of the interactions between wild American crows and dead birds challenged the idea that inappropriate mating attempts are triggered by the stimulus of copulation posture. It's not so simple. Kaeli Swift and John Marzluff found that during the breeding season, crows also tried to mate with a lifelike crow in a neutral standing position and with a crow lying on the ground in a "dead" position, its wings tucked close to its body—the latter effort accompanied by plenty of scolding. Maybe it's not the copulation position per se that inspires arousal, say the scientists. A dead posture excites alarm, which is not uncommonly followed by sexual behavior. This sex-after-alarm response has also been noted in zebra finches, vermilion flycatchers, and pied avocets. Swift and Marzluff point the finger at hormones. "It may be that breeding-related endocrine changes downregulate the ability of some birds to process conflicting information."

In any case, the literature is full of avian Davian stories, now better understood but still somewhat painful to contemplate. Among the most famous is the first documented case of homosexual necrophilia in mallard ducks.

In 1995, the director of the Natural History Museum of Rotterdam, Kees Moeliker, heard a loud *thump* against the glass facade of the museum's new wing—not an unusual event, as the glass was mirrorlike, and thrushes, pigeons, and woodcocks had frequently collided with the building in the past. "A 'bang' or a sharp 'tick' on the window meant work for the bird department," writes Moeliker. When he looked outside to investigate, he noticed a dead mallard drake lying in the sand and another one energetically relishing the whole situation. "Rather startled, I watched this scene from close quarters behind the window," wrote Moeliker. The drake's enjoyment lasted a full seventy-five minutes, and the necrophiliac mallard "only reluctantly left his 'mate,'" reported Moeliker. "I secured the dead duck and left the museum. The mallard was still present at the site, calling 'raeb-raeb' and apparently looking for his victim (who, by then, was in the freezer)."

Moeliker hypothesized that the pair were in the midst of a "rape flight attempt," when small flocks of drakes chase a single female in the air, trying to force her down to rape her. When one of the males hit the glass facade and died, Moeliker told *The Guardian*, "the other one just went for it and didn't get any negative feedback—well, didn't get any feedback."

Moeliker's research may have ruffled a few feathers, but not because of the reported male-on-male sex. Levick in his day had no idea how to interpret the penguins' same-sex sexual behavior or whether it was unique to them. Now we know it's common in the animal world. Scientists have seen same-sex sex in a wide range of species, from bonobos, sheep, and dolphins, to garter snakes, guppies, and flour beetles. Also, in more than 130 species of birds, among them, greylag geese, pukekos, cattle egrets, ruffs, and several types of gulls, including silver, California, and ring-billed. In some wild populations of western gulls, up to 15 percent of females form long-lasting same-sex bonds, performing courtship rituals, synchronized dances, and gift-giving ceremonies of bits of food, and nesting together. Researchers studying a colony of Laysan's albatrosses in Hawaii reported that fully a third of all pairs consisted of pair-bonded females that courted and coupled, groomed each other, and shared egg incubating and other parenting responsibilities.

In the past decade or two, several pairs of same-sex penguins in captivity have attracted international attention by engaging in sexual behaviors and raising chicks together. There were the male chinstrap penguins, Roy and Silo, in Manhattan's Central Park Zoo in the late 1990s, who ignored females in their enclosure, instead choosing each other and exhibiting ecstatic behavior, entwining their necks, vocalizing to each other, having sex—and, ultimately, building a nest together and attempting to incubate a round egg-like rock. After a keeper gave the pair a fertile egg to tend to, a female chick called Tango hatched, and the two raised it, warming her and feeding her until she was ready to fledge.

Some years later, two male same-sex pairs of Humboldt penguins entered the limelight. There were Z and Vielpunkt from the Bremerhaven Zoo. And Jumbs and Kermit from the Wingham Wildlife Park in Kent, England, who first got together in 2012 and, two years later, began caring for an egg given to them by park staff when the mother was abandoned by her mate. Jumbs and Kermit were so successful raising the chick together that they were deemed by the park owner "to be two of the best penguin parents we have had yet."

Scientists sometimes talk about the apparent "paradox" of same-sex sex—that it runs counter to the purported purpose of sex, procreation—and try to parse how the behavior might biologically benefit an animal.

In some birds, for instance, it might provide young males with practice, a chance to polish their courtship skills to use later with an opposite-sex partner. This was the thinking of scientists at the University of California, Irvine, when they launched a study of male budgerigars, who regularly engage in sexual activity with one another before maturity. If these males were practicing for "real" courtship with females, you would expect that males spending more time rehearsing would have better success with females later on. The study revealed just the opposite. In a paper called "Nice guys finish last," the scientists found that males that had more same-sex sex had less luck finding a female mate. Instead, the males seemed to be using these interactions—intense head bobbing, beak rubbing, and other energetic movements—to gauge their own physical condition and measure it against the vigor of other males. Budgies forage in flocks for protection—there's safety in numbers—and prefer to follow some individuals. Same-sex interactions, the theory goes, might help these intensely social birds pick their leaders.

In the case of acorn woodpeckers, male-on-male and female-on-female mounting behaviors just before the birds retire to their roosts in the evening might provide social glue for relationships, like male alliances in bottlenose dolphins, or might defuse tense situations, as it does for bonobos.

For Laysan's albatrosses, rearing an albatross chick requires the cooperation of two parents. When there's a shortage of breeding males, females will pair up. Roseate terns and California gulls do the same, all faring far better at raising chicks than lone females.

Those same-sex penguin parents became hot fodder for arguments left and right about the nature and politics of same-sex relationships. But what knowledge of these partnerships really gives us, says evolutionary biologist Marlene Zuk, is a window on what sex may mean in the bird world. "Those unfamiliar with the lives of animals in the wild often assume that sex occurs in a brisk businesslike fashion, solely for procreation," she writes. In fact, it's far more complicated than this and achieves much more. In this respect, as in so many others in bird nature, there may be versatility and variety of "purpose" in sex, same-sex or otherwise. And maybe, in some cases, the reason why such partnerships exist is simpler: They love who they love.

Chapter Ten

WILD WOOING

Clearly, there's a lot more going on in bird sex than a quick, instinctive drive to couple. But to my mind, it's the overtures to the act that most astonish. The roses-and-chocolate gestures of humans hardly hold a candle to the wild, lavish, and exuberant courtship displays of birds.

I'm writing this on Valentine's Day, and I can think of few things more romantic than the song and dance of Japanese red-crowned cranes. I saw them once in Hokkaido, tall, elegant birds raising their wings and arching their backs, bills to the sky in a primordial unison call that sent chills down my spine. As the cranes called, they began to dance, leaping together into the air, great wings flared, in a courtship ritual that practically conjugates commitment: "I have loved. I am loving. I will love."

The winter wren, a furtive little brown bird that creeps along mouselike in the leaf mold of the woodland floor, woos its mate with song alone. But what a song! A complex, rippling thread of the sweetest tones rising and falling in undulating melody. Sung with lung-bursting force—with vocal power, per unit size, ten times that of a crowing rooster—it fairly "bursts upon the ear," says naturalist Arthur Cleveland Bent, startling, entrancing, copious, rapid, prolonged, and penetrating, "at once expressive of the wildest joy and the tenderest sadness."

For heroic physical prenuptial feats, few birds outdo the male black wheatear, a small species that breeds in cliffs and stony slopes in western North Africa and Iberia. These one-and-a-half-ounce birds ferry in flight an average of four pounds of stones to waiting females in nest cavities in caves and cliffs. That's like a one-hundred-fifty-pound human carrying ten thousand pounds of stones. A single male may carry in his bill as many as eighty stones in a half-hour period. He does so in part to build a base for his nest but also to impress his mate, showing off his strength and the potential quality of his paternal care.

Some birds are solo displayers, like the winter wren or the palm cockatoo, a bizarre parrot that courts his mate with a *rat-a-tat* rhythm of love. This bird is so skilled at drumming it's known as the Ringo Starr of the bird world, with avian rock star plumage to match: a kind of goth-like smoky gray body, punctuated by flaming bare red cheek patches and a perky crest.

Robert Heinsohn first encountered the cockatoo's singular musical talents when he was doing fieldwork on the eclectus parrot at Cape York Peninsula in northern Australia. Heinsohn is founder of the Difficult Bird Research Group, a team of scientists focused on studying a handful of birds that are "extremely endangered, hard to find, occur in wild and rugged terrain, and move around the landscape." The palm cockatoo fits the bill. The bird is really a New Guinean species that has found its way into the rainforests at the tip of the peninsula. It is extremely shy and elusive, hard to find and hard to study.

"One day I heard this tapping from the edge of the rainforest and thought, 'What is that?'" recalls Heinsohn. "So I sort of snuck over and there was this beautiful palm cockatoo with a really big erect crest flushing his cheeks, and he had a stick in his foot, and he was banging the tree. He went on for about half an hour. It was incredible. After a while I worked out that there was a female there, and she was watching him the whole time."

The male starts his performance by making a tool. With his huge hooked bill, he shears off a sizable stick from a tree, snips off the foliage, and then trims it to pencil size. This in itself is a wonder. Toolmaking of any kind is rare in the natural world and almost always occurs in the context of foraging. "This is the only species other than humans we know of that makes a tool for display or for musical purposes," says Heinsohn. Holding his neatly fashioned drumstick in his left foot, the cockatoo beats it against his perch or the hollow trunk of a tree. When things really get going, he might mix in a high-pitched hooting whistle or lift his crest feathers, bob, sway, pirouette, and flash those naked cheek patches, bright red with flooded blood.

Heinsohn and his colleague Christina Zdenek spent seven years recording in the field to get one hundred thirty sequences they could analyze to try to understand the nature and function of the drumming. According to Zdenek, who did most of the recording, it worked out to capturing one drumming event in about every one hundred hours of recording. But their labors paid off. The team discovered that the cockatoos produce a regular rhythmic beat, like the beat setters of Western rock bands. And not all birds thump with the same rhythm. Individual males have their own distinctive signature cadences and percussive styles. "Some males like fast rhythm, some a slower one," says Heinsohn, "and some have this little flourish at the start of their performance." The birds always perform with their signature beat—so unique that other cockatoos may recognize a male by his sound alone.

Moreover, it appears that the habit arose through learning or culture. "While these cockatoos also live in New Guinea and Indonesia," Heinsohn says, "they've only been seen drumming on the Cape York Peninsula. Presumably one bright male hit on a behavior that pleased females, and others quickly learned it, so it spread easily through the population."

This discovery of a musical beat in the wild supports Darwin's assertion that rhythm has aesthetic appeal across species and may reflect some ancient shared aspect of brain function. As he wrote in *The Descent of Man*: "The perception, if not the enjoyment, of musical cadences and of rhythm is probably common to all animals and no doubt depends on the common physiological nature of their nervous systems."

As far as we know, there are no drum circles in the cockatoo world. The palm cockatoo's performance to win female favor is always a solo. But other male birds gather into groups to show off their wares, to collectively strut and sing and flaunt their wonderful feathers to prospective mates, most famously in leks. There's nothing like stiff competition to bring out the show-off. In fact, this communal way of courting in birds is linked with some of the most outrageous forms of mating display.

Depending on the species, a lek may be in small, congested spaces where many males converge, or more dispersed, "exploded" over larger, spaced-out display areas. Male great snipes lek together each spring in little jammed display areas in the river valleys of Russia and Poland. In a tight group, they spread their wings and flash their white tail feathers and make a glugging, whistling sound, pervasive and ventriloquial, like air bubbles released from a bog, that echoes across the dark countryside to draw in females. Some will lek tirelessly through the night, losing up to 7 percent of their body weight. Females wander through the carnival, judging the performances. They gain nothing from the venture of picking out a male but genes—no material benefits or protection, no companionship or help with parenting. Just sperm. After mating, females go off by themselves to build a nest and raise their young alone.

It might seem unwise for males to display together in clumps so dense they may be only a few body lengths apart, where competition is keen and so is the risk of harassment. But they may be drawn to these hot spots because of high female traffic or because it's the favored location of a highly successful male and they hope to intercept his females. Or because it's easier for females to find them when they're elbow to elbow. Or just because there's safety in numbers.

The ritual appears to go back to dinosaur days. Scientists working in western Colorado recently found evidence of lekking in a gigantic predatory dinosaur known as *Acrocanthosaurus*, a thirty-eight-foot ridge-backed tetrapod with feathers and crests that lived in the wetlands of western North America during the Early Cretaceous period. The team discovered fifty huge scrape marks in what is thought to have been the dinosaur's display arenas, suggesting the gigantic beasts engaged in a birdlike dance to woo mates. The scrape marks, six feet in diameter, occurred in irregular groupings that resemble the traces left behind by courting ostriches, puffins, and kakapos. Circumstantial evidence, to be sure. But fun to imagine the giant creatures lumbering about in something like the dinosaur version of avian foreplay.

Not all bird leks are alike. Some are rife with competition and conflict, like the rough-and-tumble, sometimes vicious lekking of the greater sage grouse, the fanciest bird in North America and the avian equivalent of a western gunslinger. The grouse strut around with spiked tails spread fanlike, bright white chest feathers puffed out, yellow esophageal air sacs engorged, circling each other like boxers and filling the air with strange swishing sounds. Females circulate among them, but the male who puts on the best show wins the prime position at the center of the arena, and the hens flock to him like a pop star.

"Sage grouse display side by side and they spend a lot of time knocking the crap out of each other with their wings," says Emily DuVal, an associate professor of biology at Florida State University. Males of many lekking species engage in this sort of fierce competition for mates, vigorously battling rivals. But in a handful of species, males do something highly unusual at the other end of the social spectrum: They cooperate.

Take lance-tailed manakins, birds that breed at dispersed leks in Central and South America. They have this "strange, amazing, different, and complex form of cooperation," says DuVal. To attract females, an alpha and a beta male perform tightly coordinated male-male duets and dance displays with eleven unique elements, ranging from slow butterfly flights with wing clicks and "pip flights"—quick circlings of the perch accompanied by *pip* calls at landing—to leapfrog dances, vertical bouncing, and back-and-forth hopping. "What you expect to see in lekking birds is intense competition," says DuVal, "and here you get a situation where males work together in cooperative partnerships that may last as long as six years."

But here's the thing: Only the dominant male of each pair gets to mate with the females the partners attract. Why would a male bird forgo his own mating opportunities to help another male secure his? What's in it for the wingman?

DuVal is sorting this out. She has been studying these birds at Isla Boca Brava in Panama since 1999. Her study area includes about thirty pairs of males, each presiding over a sizable display patch—about 500 to 4,500 square meters—and each with a chosen display perch. Male pairs can hear, but can't see, each other. DuVal has set up multiple video cameras to capture their courtship behavior when females visit.

At an annual meeting of animal behavior specialists, we sit on a bench in the sun while she shows me film clips of the birds from a camera in her backyard in Panama. "Just by happy coincidence, the most successful male was right off our back porch," she says.

The males are elegant in their breeding plumage, black, with a bright blue back and vibrant red crown and crest. The camera catches them on a low, bent branch, but the action starts higher up. At first, the males sit side by side in a tall tree, broadcasting their partnership with calls so precisely timed, separated by just a tenth of a second, that they sound like one bird, but in rich, resonant stereo. When a female comes into the area, the males together begin a series of coordinated whistles and calls and slow, butterfly-like flights to attract her to the display perch. As she approaches, they launch into alternating vertical leaps, popping up and down like jumping beans. If she hops onto the perch, the popping progresses to leapfrogging, each male leaping over the other and hovering just centimeters from the female's gaze. She hops rapidly back and forth to signal her interest. "Then, often with no fanfare, one male goes," says DuVal, "and the alpha is left to perform his solo slow flight back and forth in the few meters around the display perch. All of a sudden, he will swoop up to a tall perch, give a single 'pip,' then zoom back in a quick descending arc toward the female, veering up at the last second by braking suddenly with his wings, and then fluttering down from above her, his wings making a low 'whoosh' sound—which may be the point." He'll do this swooping up to ten times in a row. In between those swoops, he flies low around her with his labored slow flight, and the female appears to coordinate her back-and-forth leaps with clicks made by his wings.

Displays can last for almost three-quarters of an hour. "They can be exhausting," says DuVal. "Some males end up panting with their mouths hanging open." But this alpha is in peak physical condition. He bounces dramatically, ricocheting off the perch, and ending with a bow to the female, who lowers her wings and raises her tail slightly. Then he leaps into the air, reverses direction, and lands on her back to copulate, wings held stiffly in the air. "It took me three years to see the details of that last male bounce before mating," says DuVal. "I didn't have a camera for the first several years of the project. Now we have this front row seat." The female remains afterward for a few minutes, preening, then flits off to finish building her nest for the young she will raise alone.

"The display is more of a three-bird ballet than a simple spectator event," says DuVal. "While things often proceed with well-oiled precision, there are exceptions to the drill. It's not that there's no conflict. Cooperation doesn't necessarily mean everybody is on the same page."

In one clip she shows me, the female seems wildly enthusiastic, leaping back and forth when the males are still in the early stages of the display. She drops her wings to solicit a copulation with the beta male, who is all too happy to accommodate. But the alpha male will have none of it. He jumps on the beta male midcopulation, knocks him off the perch, and chases him away. Almost immediately, the alpha recovers and resumes his courtship display, picking up with the normal solo performance that follows a more cooperative two-male dance. The female, as she watches, defecates, which probably clears her cloaca of most of the sperm from the beta male. When the alpha returns, he takes a bow and then copulates.

I ask DuVal about the timing of the female's defecation. "Everything goes in and out the cloaca—it's one hole," she explains, "so if you have sperm that you're not going to use, one easy way to flush out the system is to defecate. After a classic display, females stretch tall and tuck their tail—it has always seemed to me that they may be rearranging themselves to hang on to the transferred sperm—and then sit around preening. Occasionally, something seems amiss, and instead of doing a stretch-and-tuck, she defecates and leaves."

The wingman is nowhere in sight. Why does he sing and dance his heart out again and again when he wins no action?

It's not that beta males are benefiting genetically by aiding a relative. DuVal has determined that they're not closely related to their alpha partners. Cooperating with an alpha male does boost their chances of later becoming alphas themselves. But it's not just about inheriting territory or waiting in line to take over if an alpha male dies. DuVal found that betas whose alpha partners were experimentally removed didn't always take over the empty alpha slot—and if they did, rarely held on to their new status from one breeding season to another. The mystery is still unfolding. "Males can live seventeen years or more, and they do not act with only short-term interests in mind but plot their career moves on an impressively long timescale," says DuVal. "One of the cool bits of this work is the cradle-to-grave approach of tracking the birds' full lifetimes in the wild, which allows us to understand how these behaviors pay off in the long term."

Birds-of-paradise are also standouts in the lekking world. In New Guinea and far northeastern Australia lives a group of forty species wildly different in their outlandish ornamental plumage—nape capes, tails three feet long, feathers fused like curlicues or plastic tabs, wirelike feathers tipped with little webbed rackets—which they use in complex, often wacky courtship displays. Not much was known about these birds until Ed Scholes of the Cornell Lab of Ornithology spent time studying them in the field and dissecting the anatomy of their displays. Some males shape-shift, says Scholes, transmogrifying from ordinary bird form into something distinctly unbirdlike, peculiar little blossoms of color or uniform circles of black. Among the most striking is the superb bird-of-paradise, which flares its ornamental cape to form a super-black oval ruff that encircles its upper body so its head vanishes into an enormous round bouncing face with blue features, like the iconic smiley face—only not smiley and a little scary. The red bird-of-paradise hangs upside down, allowing his two tail wires to fall and frame his wings in a perfect heart shape.

The species known as Carola's parotia performs an elaborate "ballerina" dance display. These birds live in the dense forests of New Guinea's mountainous interior. Their complicated courtship displays came to light when Scholes and photographer and fellow ornithologist Tim Laman observed them in Papua New Guinea. The male is black, about the size of a jay, with a bizarre set of long, wirelike head feathers that extend from behind the eyes; chin whiskers; and a dramatic mantle cape. He builds a small display court in the forest with at least one central perch spanning the length of the court, where females can sit and watch. He clears the area of debris and lays down a mat of epiphytic fungi as a dance floor. His fancy footwork begins with a horizontal perch pivot, then head tilting, back and forth, and a series of bounding hops from one end of the court to the other and back. Then comes the swaying bounce, where he wobbles or sways vigorously, wings open and fluttering, then shut tightly. When Laman photographed this move with a slow shutter and a burst of flash, the image revealed that the male swings his head in a perfect figure eight. For the iconic ballerina part of the display, the bird positions its flank plumes around its body like a tutu, then launches into an intricate dance with four phases—bow, walk, pause, and waggle—including twenty-three different dance elements.

W hy do birds have such outrageous and ornate mating displays? Darwin thought the question so vital that he devoted the bulk of his second-most-famous book to the topic. In *Descent of Man*, he argues that many elaborate displays in male animals result from the kind of evolution he called sexual selection. Females fancy males with certain traits over other males. The chosen males have more offspring, which inherit these traits, so they're passed along to the next generation and hence spread through a population. The outlandish displays we see today are echoes of female choices in eons past.

Sexual selection by female mate choice would explain the evolution of the weird dangling wattles of bellbirds and umbrellabirds, the cascading plumes of some birds-of-paradise, and the peacock's prodigal train, an ornament Darwin found so inexplicably incompatible with any contribution to the survival of its bearer that—as he wrote in a letter to Asa Gray—it "makes me sick!" Sexual selection may run counter to natural selection, driving the evolution of extravagant male courtship traits that actually hinder survival tasks such as finding food and avoiding predators. In this way, males may end up with elaborate traits and displays that are burdensome but so sexy they're perpetuated in the population.

The engine of nature's diverse and exuberant sexual displays, Darwin argued, is female preference. "If man can in a short time give beauty and an elegant carriage to his bantams, according to his standard of beauty," he wrote, "I can see no good reason to doubt that female birds, by selecting, during thousands of generations, the most melodious or beautiful males, according to their standard of beauty, might produce a marked effect."

Darwin's thesis was elegant, but it was met with cold reception. His contemporaries expressed deep skepticism over the idea that a "taste for the beautiful" could explain so much spectacle in nature—and also, downright disbelief that an animal, particularly a female animal, could have an aesthetic sense sufficiently sophisticated to make fine distinctions. Alfred Russel Wallace argued that male ornaments and displays instead arose from a "surplus of strength, vitality, and growth-power which is able to expend itself in this way without injury." Courtship and its displays were controlled by males and male competition, and the primary purpose of exuberant display was to give males the advantage in competition or conflict with other males.

The idea that a peacock fanning its flamboyant tail or a bird-of-paradise hanging upside down in a supremely vulnerable position might be trying to intimidate another male rather than displaying his splendor to a female may seem absurd. But for decades after the publication of *Descent of Man*, Wallace's viewpoint prevailed. "Seldom or never does the female exert any choice," wrote philosopher and psychologist Karl Groos in 1898. "She is not awarder of the prize, but rather a hunted creature."

Now we know that the female is in fact most often the hunter, the arbiter of mates, and her choice drives the evolution of male sexual display. But we're still striving to understand just how. What are females looking for? And how do they distinguish a first-rate show from a flop?

These are not easy questions to answer. What matters is the bird's perspective, not ours, and it's often difficult to get a bird's-eye view. For example, plumage and other body features used for sexual display may look dull to our eye, but add ultraviolet perception into the mix, and it's a different story. In a classic study in the late 1990s, biologists applied sunblock which absorbed UV light to the feathers of male blue tits and discovered that this affected how attractive they were to females. Likewise, scientists recently discovered that the namesake horn of the rhinoceros auklet—a protuberance grown in preparation for mate selection—apparently colorless to us, actually glows in ultraviolet light, popping out to the female eye as a sexual signal.

We humans are constrained not just by our limited senses but also by our perception of time. In the bird world, things happen fast, sometimes too fast for us to see. To make this point in talks, Mike Webster of the Cornell Lab of Ornithology shows a real-time video made by biologist Lainy Day of a male black manakin displaying in the forests of Guyana. In the film, the male manakin looks like it's simply hopping up and down. Then Webster plays Day's high-speed video, which shows hundreds of frames per second, as the female manakin would see it. Jaws drop, and there's an audible gasp from the audience. Between the little hops, the male completes a full-body, 360-degree flip, a high-speed somersault too quick for us to see.

The wonders of high-speed video have also revealed the sensational tricks performed by blue-capped cordon-bleus, songbirds native to Africa. Both males and females perform rapid foot movements while bobbing and singing. When Nao Ota of Hokkaido University filmed the birds at three hundred frames per second, she discovered that the birds tap-dance in perfect time with their song.

What do females actually experience when they view a courting male? And what criteria do they use to pick their mates? These are questions that obsess Mary Caswell Stoddard, who studies the extraordinary mating displays of the broad-tailed hummingbird.

Consider what it takes for a male of this species to court a mate. Jet up nearly vertically like a little firework, whirling your wings forty times per second until you reach a magic spot up to a hundred feet in the air. Stop midair, hover for a few seconds, and then, with a burst of wingbeats, execute a power dive, plunging rapidly down, down, down at death-defying speed toward your target—a female perched somewhere below the U-shaped nadir of your dive. When you reach her going full speed, dazzle her with your throat feathers at just the right angle to catch the light. Then swoop back up the other side, up, up, up on your green wheel of wings until you reach that magic height. Then dive down again in the opposite direction. Repeat as often as your little body will allow—sometimes dozens of dives per hour.

It's "one of the world's coolest, weirdest courtship displays," says Stoddard. The display gets even crazier when you analyze its details, which is what Stoddard and her colleague Benedict Hogan have done in the rarefied air of Gothic, Colorado, the site of the Rocky Mountain Biological Laboratory, where around three hundred broad-tailed hummingbirds come to breed, having spent the winter months in Central America. Most of what we know about these birds derives from studies conducted at RMBL (affectionately known as "Rumble"), one of the biggest, oldest biological field stations in the world, located at over a vertical mile of elevation in a former ghost town from Colorado's days of silver mining in the nineteenth century. We watched the hummingbirds outside a research hut in a mountain meadow blooming with Colorado columbines, lupines, bluebells, and scarlet gilias.

Stoddard studies how evolution shapes the diversity of color, pattern, and structure in the natural world. Here at RMBL, she and her colleagues—Hogan, Harold Eyster, and David Inouye—are exploring how hummingbirds use color to find flowers and woo mates in the wild. At the moment, the team is investigating the broad-tailed hummingbird's ability to distinguish between colors that differ only slightly. A pair of specially designed nectar feeders sit on tripods in front of custommade LED devices that beam one of two colors. Only one hue is associated with a sugary reward, so the birds must choose wisely. Often, the two colors, such as green and ultraviolet green, are completely indistinguishable to the human eye, but the birds learn quickly, over the course of a day, to make the distinctions. "It's magic to see this expression of their different visual environment," says Eyster.

You hear the hummingbirds before you see them, a shrill metallic-sounding trill made by their wings. They're no bigger than my thumb but with a reputation for boldness, intelligence, acrobatics, and sheer beauty. When the Spanish first arrived in the Americas, they described the birds as *joyas voladoras*, or "flying jewels." *Boom!* One shoots up to my face like a great bee, stops short and hovers before me as if I were a flower, flashing its rose-magenta gorget, like a chip of brilliance from a volcanic sunset. Then *poof!* It's gone.

The broad-tailed belongs to a tribe of bee hummingbirds known for their sensational diving displays. It's a promiscuous breeder, each male often mating with several females in a single breeding season. The denouement is rarely witnessed by the human eye. No pair-bond is formed. After mating, the female will build her tiny nest from spiderweb and gossamer, twisting it around and around her body until it forms a cup. Then she will tend it alone, and alone raise her brood.

Scientists have known for some time that these hummingbirds execute wild dives during their displays, that they flash their brilliant gorgets, that they make special sounds. Christopher Clark, now a biologist at the University of California, Riverside, discovered that these sounds are not vocalizations but sonations, mechanical noises the birds make as air passes over specially modified feathers in their wings or their tails and vibrates them. The birds are probably capable of making some of these sounds during normal flight, but the tail-generated buzz they reserve only for their dives.

What remained a mystery was how all of these elements come together and what the female might see as she perches in a shrub just below the base of the dive. This is vital information if you want to understand these displays and how they came to be, says Stoddard. You have to understand the female's perspective. "But because the displays are dynamic and operating in three different domains—motion, sound, color—they're difficult to study," she says. "We had to find the right tools for quantifying how the complex signals are produced together and how they would be perceived by the female."

First, Stoddard and Hogan made video and audio recordings of forty-eight dives using a GoPro camera set up near their research hut and, with image-tracking software, figured out the trajectory and speed of each dive. Then they used acoustic analysis to map sound onto the dive to measure the exact moment when the males make that strange buzzing snap with their tail feathers—Stoddard describes it as a little *nyeh*, *nyeh*—and estimated what it would sound like to the female as it whirs past her.

Figuring out what was going on with the bird's flashing gorget was tricky. The iridescent throat of the hummingbird is made of feathers that are structurally colored, so their appearance shifts dramatically from magenta to black depending on the female's angle of view and the illumination. Male Anna's hummingbirds know this trick and dive toward the sun to flash their red gorget at a female spectator, maximizing its showiness. As one early observer wrote, "The effect is one of a tiny ember suddenly descending upon the observer, growing in brilliance and dimension as it approaches, to burst with a pop as it passes over the display object."

To try to understand how the female broad-tailed hummingbird might see the gorget's shifting color through the dive display, Stoddard and Hogan turned to an old-fashioned cache of information: a collection of hummingbird skins at the American Museum of Natural History in New York. "We built a special 3-D-printed rotating stage for the bird skins," says Hogan, "and then photographed the specimens from different angles using an ultraviolet-sensitive camera to capture the four-color-cone vision of hummingbirds." To estimate how the colors might appear to a bird with its capacity to see UV light, they used a suite of software tools, including Stoddard's TetraColorSpace computer program. By combining the photographs with their tracking of the male's body position on his U-shaped flight path, they could estimate the female's view of the gorget.

When they put it all together—motion, sound, gorget flash—the pyrotechnics of the dive unfolded. As a male descends, diving beak-first, he goes faster and faster until, at the bottom of the U-curve, when he's nearest the female, he hits top speed. According to Christopher Clark, who studies the similar dives of Anna's hummingbirds, the average maximum speed of a displaying hummingbird is around 385 body lengths per second, almost twice the speed of a peregrine falcon diving in pursuit of prey or a swallow diving from high altitude, and—relatively speaking—greater than the top speed of a fighter jet with its afterburners on (150 body lengths per second) or the space shuttle during atmospheric reentry (207 body lengths per second). When the bird pulls out of the swoop to avoid crashing, it experiences g-forces up to ten times the force of gravity. In that range a human jet fighter pilot may temporarily go blind or black out. Not the hummingbird. Going at full throttle at the bottom of his U, he starts his tail buzz. As he approaches, the female hears an upward shift in the pitch of the buzz and, as he departs, a downward shift. This is the Doppler effect, the same phenomenon that creates the changing blare of a horn as a car moves past you and tells you whether the car is zooming by or just cruising. The sound gives her information on his speed. As he enters her field of vision, his gorget flashes a vivid, sexy magenta, then shifts dramatically—in a mere 120 milliseconds—from bright red to dark green, almost black, as he passes over her.

In other words, at the bottom of his dive, when he's closest to the female, the male pulls out all the stops. "He's synchronizing these key events—peak speed, the sound of his feather buzz, and the dynamic flash of his bright iridescent gorget—so they happen all at once just as he passes over the female," says Stoddard, "producing one big in-your-face sensory explosion of signals to really wow her."

Like all good science, Stoddard and Hogan's research raises more questions than it answers.

What components of the male's display really matter to the female? Is it his speed? His buzz? His glowing gorget or its rapid color shift from red to black? And what part of her mind can gauge the subtle differences between one dive-bombing male and another?

"We don't know," says Stoddard. "Maybe it's not just that the male is going fast," she muses. "And maybe it's not just that he has this amazing mechanically produced sound or that he's colorful. Maybe what really matters is his tight synchronization of all three of his tricks." However, it looks to the researchers like there's not much variation in the degree of synchrony between the dives of different males; they're all pretty much spot-on in their timing. "If all the male dives seem similar to our eyes," says Stoddard, "what fine distinctions are we missing?" She wonders if it might be the accuracy of a male's aim, his ability to deliver his show straight to her eye as she sits directly beneath him.

The point is, it's the bird's-eye view that matters—the female's view, to be specific. She's the one gauging the male's performance and deciding if he's worthy as a mate. And in so doing, she's shaping the courtship behavior—the crazy flirtatious rituals—of her whole race.

U nderstanding a female's experience and her preferences may shed
light on how this sort of complex and bizarre display evolved in the
first place.

There are a few theories about this, says Stoddard. The first is that
females are deriving information about the male's quality from the various aspects of his display. "One possibility is that males that go faster,
shine brighter, and are better synchronized are genetically healthier
males," says Stoddard. This has been the dominant view of much of animal signaling for centuries. And for many scientists, it's the most compelling explanation for the appeal of extravagant traits or displays: The
male with the best show has the best genes.

Even simple displays can offer females information about male quality. A female Adélie penguin, for instance, can tell from a male's croaky,
squawking song how fat he is and thus how good a father he will be. Successfully raising a family in Antarctica requires the devoted efforts of
two penguin parents. Male and female take turns incubating their eggs
and watching over the chicks while the other parent searches for food. As
Levick noted, males arrive at breeding grounds first, claim their territory, and build a nest. When females arrive, they want to know how long
a male will be able to brood the eggs without having to meander off in
search of a meal. A male shows off his bulk by throwing back his head
and letting loose with a wild trilling squawk like a braying donkey. Scientists have recently found that the fat around his voice box affects the con-
sistency of pitch and steadiness of his call. Females listen intently for this
glottal giveaway and snatch up the pudgier males.

From the complex melodious song of a male winter wren, a female
may get reliable news of his vigor and condition, including the quality of
his brain and its development. Male songbirds in poor condition or that
experienced stress when they were developing, such as disease or too little food or too much sibling competition, don't develop healthy neural

song systems and can't produce high-quality attractive songs. Females pick a mate based on subtle variations in the acoustic features of his song, which are reliable signals of his vigor and genetic health.

Or consider what's divulged in the acrobatic display of a golden-collared manakin. The male's wing-snap courtship trick, which he produces by rapidly hitting his wings together over his back so they meet with that audible snap, relies on superfast muscle contractions. The manakin can snap its wings together more than one hundred times per second. And he'll do this snapping on the go, while he's leaping from perch to perch, and then finish his show with a kind of half flip jump to the ground, in a landing so perfect it would shame an Olympic gymnast. Matthew Fuxjager of Brown University has found that females mate with males that display at split-second faster speeds. Remarkably, the females can detect such minute differences.

These kinds of physical displays—leaping, diving, wing clapping, simultaneous singing and dancing—reveal a male's vigor, his ability to perform energetic acts over and over again. Vigor can't be faked; it's considered an "honest" signal of male condition, so choosy females pick males with the most vigorous motor performance. In the speed of a male broad-tailed hummingbird's dives and how he flashes his gorget, a female may judge his physical prowess. "She may not just be interested in how red his throat is, but how quickly it turns from red to black," says Stoddard, "which might give some indication of the geometry and speed of the dive, which may be linked to male vigor."

From elaborate displays, then, females may glean information about male fitness and quality, his "good genes."

That's one idea, says Stoddard, but there are others. The "signal efficiency" theory, for instance, which might be called the bright beacon hypothesis: Complex signals may have evolved not because they're conveying information about quality, but because they efficiently communicate, "Here's a male who is ready to mate!" The more signals you make, and the brighter, louder, and flashier they are, the more conspicuous you will be to members of the opposite sex. Some scientists argue that females of certain bird species are drawn to these "supernormal stimuli" in males—bigger, brighter, louder traits—simply because that makes them easier to spot. Females who can quickly locate a male are less likely to be eaten by a predator and can spend their time and energy on foraging and finding a nest site.

Having more than one mode of display—sound, sight, motion—and synchronizing them may also help the female with so-called sensory integration, binding the sensory input from multiple senses into a coherent whole. Not unlike the way humans may require synchronized lip movement and voice to understand speech.

In addition, the more complex the signal, the more likely it is to hold the attention of a female. If a simple signal is repeated over and over, she may become habituated to it, and her interest may wane. In general, birds pay less attention to constant noise than to the sporadic or unexpected sound of a hawk call or a branch cracking beneath the paw of a predator.

Plus, having both a visual and an audio component means that you have backup under difficult conditions, says Stoddard. "On a noisy day, you still have your flashing gorget. On an overcast day when you don't have a lot of sun, your gorget might not be as striking, but your tail-generated buzz will still be audible. Those signals together ensure that the female takes notice—even if comparisons between one male and another, who is brighter or faster, don't offer a lot of information."

These two explanations—the "good genes" and the "bright beacons"—have in common the view that ornate displays have evolved because they're useful—they signify quality or draw attention to a courting male.

But there's another perspective, says Stoddard, more controversial but deeply intriguing. The hummingbird's flamboyant dive may have evolved for a more capricious reason: "Simply because it's beautiful!" she exclaims. "And females love it, and that's enough to create selection pressure for aesthetically pleasing things."

Maybe, as Emerson said, beauty is its own excuse for being.

The notion that complex displays have evolved for purely aesthetic reasons—not because they're objectively informative but because they're subjectively pleasing to the chooser—is an old one, dating back to Darwin and developed by geneticist Ronald Fisher a century ago. Over generations, said Fisher, extreme traits and the desire for them could coevolve, leading to "Fisherian runaway selection." Two scientists have recently revived these ideas, Michael Ryan, an evolutionary biologist at the University of Texas at Austin, and Richard Prum, an ornithologist at Yale. In his book *The Evolution of Beauty*, Prum revisits Darwin's idea and rebrands it the "beauty happens" hypothesis. Females develop an arbitrary fondness for certain aesthetic male traits, even if they aren't informative—an acrobatic dive, a set of flamboyant feathers, a fluorescent bill, a buzzing tail feather. Their offspring inherit the traits they consider beautiful and sexy, as well as the penchant for them, and then these traits spread through a population through runaway selection.

In other words, male broad-tailed hummingbirds dive, flash, and buzz not to show off their genetic quality, but because females like what they see.

"When Ben and I discussed the broad-tail's diving display with Rick," says Stoddard, "he said he believed that the stunning dive—with its tight synchrony and mash-up of multiple sensory features—could certainly have evolved simply because females found it beautiful and sexy. Full stop.

"Which of these hypotheses is true? We don't know. It may be a combination of all three," says Stoddard.

One way these theories might work together has been suggested by evolutionary biologist Marlene Zuk. When you look at an elaborate trait, says Zuk, whether it's a peacock's tail or a difficult diving display, and ask, "What's going on with that?" you are really asking several questions. How did the trait get there? Why that trait and not some other? And how is it maintained in a population? A flashy trait might attract a female initially because of her sensory bias and then become more common through runaway selection. Perhaps the flashy trait becomes linked to genes critical for health, says Zuk, for parasite resistance, say—Zuk's favorite idea—and hence indicates a higher quality in males. It's a kind of marriage of the bright beacon hypothesis and the good gene theory.

Whatever the reason for the evolution of dazzling courtship displays in birds, they often have one thing in common—they require sophisticated brainpower. The long song repertoires of the winter wren, the coordinated displays of hummingbirds, the collaborative songs and acrobatics of lance-tailed manakins are all behaviors demanding motor skills coordinated by the brain. Are complex mating displays also a way of showing off intelligence in birds? Do choosy females choose brainy mates?

Some answers are emerging from a family of birds that takes courting to extremes. Male bowerbirds not only exhibit some of the most intricate spectacles of any species, involving building, decorating, dancing, and singing, they also manipulate how females experience their dazzling show.

Chapter Eleven

BRAIN TEASERS

One September morning I sat on a stoop in the tiny Australian town of Baradine and watched a male spotted bowerbird courting a female in his bower just ten yards from where I sat. The bower was well camouflaged beneath some shrubs in someone's backyard, and the birds seemed unfazed by my presence. At first glance, the spotted bowerbird is not much to look at: a slim, compact bird about the size of a common grackle, with dull brown plumage and a stroke of pale cream on its belly and a little lilac crown, subtle when it's not flared. But what it lacks for in plumage pizzazz it more than makes up for in flamboyant behavioral display.

With me were a few members of the Australian Wildlife Sound Recording Group hoping to capture the bird's parade of mimicked calls and songs. At its start, though, the morning's show was largely silent.

Silent but still spectacular. Bowerbirds are lekking birds, like manakins, birds-of-paradise, sage grouse, and hummingbirds, with leks dispersed over wide territory. To attract females, males build bowers, ornate structures made of sticks and decorated with ornamental objects. These are not nests; no raising of young takes place here. Rather they are theaters for seduction, the stage a male bowerbird uses as a backdrop for his song and dance to woo visiting females. Of the nineteen species of bowerbirds, fifteen build bowers, each favoring a different structure and different sorts of decorations, each remarkable in its own way. Macgregor's bowerbirds build a maypole spire of twigs and sticks up to three feet high and circled with moss and piles of insects, nuts, and fruit. Male satin bowerbirds collect blue objects and paint the interior of their bowers, chewing up dried hoop pine needles and applying the resulting brown paste to the inside of their bower walls. Only humans and a few other animals create more lavishly decorated structures. When Jared Diamond ran across the beautifully woven child-size hut of a Vogelkop bowerbird in New Guinea, sumptuously bedecked with hundreds of blossoms, berries, and other natural objects, the evolutionary biologist called bowerbirds the "most intriguingly human of birds."

Spotted bowerbirds have a well-earned reputation for being extreme in a family of extreme birds. They treasure all variety of objects, among them shells and bones, of sheep, emu—whatever might be scattered about. Ornithologist Alec Chisholm noted that one bower in New South Wales held more than 1,300 bones. "Imagine the strength of the decorative impulse that caused a bird to carry in its beak, one by one, and over long distances, that large number of sturdy bones," wrote Chisholm. He also marveled at the 2,500 snail shells he discovered in and near another bird's bower, "each somewhat bulky relative to the size of the bird's beak," and nonetheless carried over considerable distances.

This male's bower was a showpiece, two walls built not of dense sticks like those of its more famous cousin, the satin bowerbird, but of thin strawlike grass, creating a fine little corridor hedged by see-through walls, where the female takes refuge during courtship. The display area glistened with dozens of shards of green glass, screws, flip-tops from cans, bits of broken jewelry, straws, stones, seedpods from various native trees, and a few stray shreds of red ribbon, plastic, and wire. Many of the objects were sorted by color into little piles of green and white. Nested in the bowl of the bower, where the female stood, was a magnificent collection of clear glass marbles that sparkled in the sunlight.

The male entered from the left and pogoed wildly from one side of his bower to the other, hopping in and out of the sunlight, sometimes careening into his own wall. He picked up a bit of red ribbon in his beak and pranced around with it, then flung it high, completely out of his bower onto an overhanging twig. He flared his little pink crest with dramatic effect. He lowered his head like a ram and flicked his wings back, juddering from head to tail. Then, suddenly—to the delight of my companions—he launched into full vocal mode, buzzing, rasping, hissing, chirring, *skraa*-ing, twanging, cawing, gurgling, twittering, all in time with his dance moves. Some of his calls were grating, metallic, and jarring; others musical, whistled notes. Like the superb lyrebird, the spotted bowerbird is a master mimic, known to imitate the squealing of rabbits, meowing of cats, barking of dogs, rasping of branches, even the rolling of thunder—as well as numerous bird voices, including the keening cry of a whistling kite.

The female, meanwhile, kept her distance, hopping around the display area, but always keeping one wall of the bower between her and her would-be mate, peering through the fine grass to watch him from a safe vantage point. Bowerbird males display in exploded leks. Female bowerbirds, like females in other lekking species, receive only sperm from males, no help with nest building or parenting. They gauge a male's suitability by his bower and his show. Male spotted bowerbirds with well-built bowers win more matings than those with sloppier bowers, so the design of a male's bower and the tenor of his show are carefully crafted and designed to influence what the female sees. Scientists spending long, laborious hours observing, filming, and recording the courtship of a range of bowerbird species have discovered just how shrewd and tactical these birds can be in shaping a female's experience.

According to Gerald Borgia, who has studied bowerbirds for more than three decades, and Jason Keagy, an evolutionary biologist at the University of Illinois at Urbana-Champaign, male spotted bowerbirds strategically place decorations of different colors and sizes to construct a series of successive scenes for the female to see as she approaches and enters the bower. In the dim shade beneath trees or shrubs, where a male usually builds his bower, it helps to have a beacon to catch a female's eye and alert her to its presence, so on the perimeter of his bower, the bird places big visible decorations like bones that act as long-distance signals to draw females in. Inside his little landscaped garden, he creates a neat, narrow lane clear of big decorations, sometimes carpeted with flat stones or sticks, where he will do his song and dance display. Next to the path, he may stack bones into piles, forming a white backdrop for his bright pink crest. "When we placed vertebrae or other large decoration on the path," say the scientists, "they were quickly cleared away." Once the female has hopped into his little straw bower, she encounters his most highly prized possessions, an array of glistening gems carefully collected in the bower's central bowl to stimulate her.

Males often traverse the route traveled by females into the bower, say the scientists. "This allows them to view scenes in the same way females do and to build and adjust them to match female preferences at each step." Once the female is in the heart of the bower, that narrow space between its walls dictates how she stands and what she sees as the male courts her. Borgia argues that, over time, male spotted bowerbirds may have modified the walls of their bowers to make a female feel safer, less at risk. "An intense, aggressive male display is attractive to females because it provides information about overall male vigor, and ultimately, genetic quality," he says. But it can also be threatening. Those thin bower walls may function as a filter, "allowing the female to view aggressive high-energy male displays from a protected position."

The great bowerbird, a species that lives in northwestern Australia, takes even greater pains to manipulate the female's perception of his glories. These birds build bowers with an avenue up to three feet long hedged by two parallel densely thatched stick walls, open at both ends to a display court with carefully arranged objects. Males place together pale objects like stones, bones, and bleached shells—sometimes hundreds of them—to create a plain background for more colorful objects, like green sticks or red fruit. Laura Kelley and John Endler have found that the birds carefully organize these pale objects by size. That in itself is not so surprising. After all, in building its little cone-shaped home, the lowly trumpet worm selects and sorts sand grains and fits them together precisely so the smaller ones lie at the narrow end and the heftier ones at the wide end. But the bowerbird does the worm one better. He arranges his objects to create an ingenious forced-perspective illusion, which is thought to make both himself and his colorful objects look bigger to the female when she watches from the bower. In creating this pattern, the birds use the same technique humans use in laying mosaics: Start in the middle—the center of the court—and work outward in both directions. When scientists destroyed the illusion by placing the smallest objects in front and the largest in back, the bowerbird quickly rearranged the stones to restore it. For good reason. The better the illusion, the better the bird's chances of mating.

There's more. When a male great bowerbird displays, he stands at the edge of an avenue entrance so just his head shows to the female in the avenue, then plucks up a decoration and parades it in front of her before flinging it across the court in favor of a new decorative object. The color pops against the monotone gray-white background of the bones and stones. On average, he flashes five objects per display, alternating these with a quick flare of his magenta crest. And here's the astounding thing— males boost the impact of the colors in their show with a special technique: They build the interior of their avenues with red sticks that create a reddish light within the bower, actually changing the way the female sees. Exposure to this reddish light for a minute or more results in chromatic adaptation in the female's eyes, altering her perception of the colored object he displays. As a result, she experiences stronger color sensations, especially with certain hues, like the magenta of the male's crest. He also uses the structure of his bower to hide objects from the female and then flashes them to grab her notice. Females stay at the bower longer when a male presents objects with a variety of colors and lots of changes between them during his display. Quick flashes of multiple colors hold her attention. Moreover, he'll alternate between high-contrast colored objects, like green fruit and his own pink crest, with low-contrast objects like brown sticks, effectively maximizing the novelty and surprise in his display—like a male lyrebird lacing his repertoire with a variety of mimicked sounds to capture and hold a female's fancy.

I was entranced by this male spotted bowerbird's performance—and so, apparently, was the female. She hung around for a good fifteen minutes or so.

I'm guessing he was an old hand at this. Experience counts in the world of the bowerbird. It can take the better part of a lifetime to produce a really good display. Young males appear to learn the finer points of the craft from their elders, making regular visits to active bowers during their long adolescence and watching adults perform their pageants. This takes brains. Young males have to learn and remember the elements of a winning display—what novel decoration or bit of mimicry works to enchant and stimulate a female and what turns her off. For instance, in composing their jazzy high-intensity song and dance, experienced males may take into consideration the sensitivities of the female they're courting and modulate their performance to lessen the threat, alternating between forceful and softer elements. Young male bowerbirds learn these tricks.

In this way, bower design and display style may actually be culturally transmitted, like the song dialects of birds and human arts and customs. Joah Madden of the University of Exeter and his colleagues have good evidence of this. They have found that within an area, different populations of the same bowerbird species diverge in their bower design and decorations. At one park site in Queensland, Madden found that all of the bowers in a particular geographic location used similar decorations, and not just because those were the decorations available. The bowerbirds at the site had equal access to most types of decorations—mussel shells, white quartz, emu eggshells, red and black plastic, blue glass. But birds in the north seemed to favor white quartz; those in the southeast, red and black plastic and metal decorations; in the west, it was blue and purple glass.

Because a male's bower and his particular song and dance performance bear his signature, an adolescent male learning at the feet of an
elder may carry on his tradition.

In the end, only a tiny percentage of displays—even from practiced
adults—win a mating. Borgia found that just 53 of 1,284 courtships he
observed among spotted bowerbirds resulted in copulation. Females are
very choosy. They visit the bowers of different males often, repeatedly
sampling their designs and displays until they pick a mate.

What charms her? The vigor and sensitivity of his performance, the
quality of his bower construction—the amount of thin grasses used, the
sculpturing, symmetry, and verticality of its walls—and even more important, the number of bone and glass decorations. All of this she judges,
comparing it in her mind with the talents of other males she has visited—
spotting the differences in decoration quality and numbers and recalling
these differences as she moves between bowers.

She may not just be counting decorations or sizing up constructive
prowess or facility in dance or song, says Gerald Borgia. She may be selecting a mate for something that encompasses and enables all of these
facilities. Bones and glass don't signify to her different kinds of quality in
a male, says Borgia, but rather his overall tactical strategy, which she may
recognize as a sign of his brainpower. Likewise, female great bowerbirds
may use a male's ability to find appropriate objects and transport them to
his bower, as well as the quality of his forced-perspective illusion, as
marks of his intelligence and ability to learn from experience—ultimately
using these traits to gauge his value as a mate.

Darwin proposed that clever individuals are preferred as mating
partners, so mate choice might contribute to the evolution of cognitive abilities as much as to ornaments and song. A "smart" partner can
offer a female direct benefits—such as better food extraction and the

capacity to provide superior shelter in the face of varying environmental conditions, which may boost her chances of survival and the number of her offspring—or indirect benefits in the form of progeny with better cognition, which may contribute to their health, longevity, and reproductive success.

But mate choice for cognition is not an easy thing to prove. In the case of bowerbirds, researchers may infer that females prefer mates with cognitive superiority based on secondary behaviors linked with intelligence—a male's fancy bower, song, and dance. But this is only a correlation. Choosy females are not directly observing the cognitive performance of potential mates.

An ingenious new study offers more direct evidence that at least some female birds actually favor smarts in their mates. In 2019, Jiani Chen, a researcher at the Chinese Academy of Sciences' Institute of Zoology, and her colleagues tested the concept in budgerigars, those little Australian parakeets commonly kept as pets. (The name is said to come from the native Australian indigenous word *betcherrygah*—*betcherry* meaning "good," and *gah* meaning "parakeet.")

I grew up with a budgie, a spunky little male named Gre-Gre, who showed off his problem-solving prowess every morning by navigating the maze of cereal boxes I placed around my cereal bowl to prevent him from perching on the edge and pilfering my wheat flakes. My efforts invariably failed.

Like most parrots, these birds are clever—by necessity. Budgies are native to Australia's arid outback, where food sources such as insects and seeds are often unpredictable and challenging to find and extract. To survive, a parakeet must have the cognitive skills to solve foraging problems, or a mate who can do so. Unlike bowerbirds, female budgies depend on mates to provide food for them while they incubate, brood, and feed their young, so acquiring a partner good at the trick of finding food is a big advantage.

In the experiment, each female was offered the companionship of two males to see which she preferred. Then the researchers spent a week training the "less preferred" male on a foraging puzzle—teaching him to open a tricky translucent container full of seed. Next, each female was allowed to watch the trained (but previously scorned) male handily solve the puzzle again and again, and in between, to watch the untrained (but previously favored) male be completely stumped by the task. The females had in their compartments their own seed-filled puzzle boxes taped shut so they were aware of the difficulty of the task at hand. After the observation period came the moment of reckoning. The females were once again asked to choose between the two males. Remarkably, they all shifted their preference to the males they had once rebuffed but that had shown themselves to be adept problem solvers.

This would seem to be a clear-cut example of the "smart is sexy" idea. While there are alternative explanations—the females might interpret a male's ability to open a container as a show of superior physical strength, for instance—it makes sense that clever behavior in a foraging task would appeal to female budgies and that directly observing this and finding it attractive could contribute to the evolution of cognitive abilities underlying these kinds of skills.

If a male uses impressive brainpower to fashion a display or solve a foraging problem, the female equals this in her ability to assess his efforts. After all, she has shaped his show.

When it comes to preferences and choosiness, Michael Ryan reminds us, "the brain is where the real action is." Whether a female bird picks her mate for his spectacular plumage, elaborate song and dance, or acrobatic prowess or mental agility, the seat of her judgment resides in her brain—which has evolved strong discrimination abilities that allow her to readily evaluate male quality and distinguish between potential mates. Think of the subtleties of this. A female broad-tailed hummingbird can parse complex, multifaceted dives that look to us nearly identical. Female golden-collared manakins distinguish split-second differences in a male's acrobatic display. A female spotted bowerbird assesses the architecture, decorations, and performance of numerous males and holds these comparisons in her head.

As awestruck as we may be by a male bowerbird's ingenuity, we should be equally astonished by the female's ability to judge them, how sensitive her perceptions must be, how refined her aesthetics and how sophisticated her powers of discrimination, to push males to such extremes of behavior, aesthetics, and intelligence, to win her favor.

And once the choice is made? Is it made only once? Are birds loyal to their mates?

Not by a long shot.

It is now well known that sexual monogamy in the bird world—once considered the predominant way of mating—is largely a myth. The old view dates back to the 1960s, when British evolutionary biologist David Lack assembled information on what was then known about mating systems in birds and concluded that more than 90 percent of species were monogamous. Two decades later, DNA fingerprinting turned the field on its head, revealing that most birds, even most socially monogamous birds, were "playing away" a good deal. Nests of paired birds frequently contained sibling chicks with fathers other than the males that provisioned and defended them.

The revelation illuminated the complexity and diversity of mating systems in birds and put to rest the romantic notion of devoted and monogamous avian mates—to the point where exclusive monogamy in a bird pair is actually considered noteworthy. Among the handful of true monogamists are mute swans, black vultures, scarlet macaws, bald eagles, Laysan's albatrosses, whooping cranes, California condors, and Atlantic puffins. At the other end of the spectrum are the champions of extra-pair sex, Australian magpies and fairy-wrens. The mismatch between the fairy-wrens' social mating system—the birds form lifelong pair-bonds—and their sexual mating system is nothing short of spectacular. Studies have shown that up to two-thirds of offspring in fairy-wren nests are sired by extra-pair males—the highest known incidence of cuckoldry in birds.

At first it was assumed that male birds were the primary perpetrators of all this infidelity, sneaking off for "extra-pair copulations" with other females, disseminating their sperm—and their genes—as widely as possible. Females were thought to be the victims, suffering both the indignity of sharing the attentions of their social partner with other females and, worse, as British biologist Tim Birkhead put it, subject to "forced extra-pair copulation."

Views changed in the 1990s, when scientists started radio tracking female birds, from fairy-wrens to hooded warblers, and discovered that the females were doing much of the active soliciting.

Evolutionary ecologist Andrew Cockburn has been studying superb fairy-wrens at the Australian National Botanic Gardens since 1987. He has a long history of working in weird mating systems. He began his career investigating *Antechinus*, a group of small, mouse-like marsupials that exhibit the strange phenomenon of semelparity. Soon after mating, males die. But after years studying the bizarre mammals, he says, he realized that he "was not having nearly as much fun tramping through leech-infested rainforest as one of [his] graduate students was having teasing apart the intricate sex lives of superb fairy-wrens in the croissant-infested Botanic Gardens in Canberra."

In the past decade or so, Cockburn and his colleagues have discovered that female fairy-wrens pretty much run the cuckoldry show, foraying out for philandering before the bakeries open. In the few days before egg laying, and in the dark hours before dawn, females leave the bush, where they roost with their social group, and head for the territory of preferred males. These are males who, for months before the breeding season, have been leaving their territories and displaying to females living as far as seven territories from their own. Often, males enhance these displays with a trill song and one of those seemingly sweet courtly flower presentations. After observing a female from cover nearby, a male approaches her with a curled yellow flower petal carried in his beak and assumes a pose, erecting his cheek and crown feathers, lowering his tail, and displaying the brilliant blue and black feathers of his plumage by twisting his body from one side to the other to maximize the contrast with the yellow petal. Flower-petal presentations, then, are not about finding a social mate, but about finding sex with someone else's mate. When the time comes, males let foraging females know where they are, and advertise their availability by singing during the dawn chorus. Only the early birds get the flirt.

Whether males and females wander has a lot to do with how they raise their young. The mélange of mating systems in birds reflects their wildly different styles of parenting.

Parent

Chapter Twelve

FREE-RANGE
PARENTING

When the phone call came from Tim Low, I was waiting at the Brisbane station in Queensland, about to board a train to his home in Indooroopilly, a suburb four miles west of the city. Low's voice was muted, almost a whisper: "It's here." Great, I said, I'll get there as soon as I can. It was a twelve-minute train ride, but then Low and I had trouble finding each other at the station, and by the time we reached the house, it had vanished. Low said it wouldn't be gone for long, so we bided our time observing the possums living in the rafters of his garage, long tails hanging down like streamers. Parts of Indooroopilly were once rainforest, and there are still remnant trees, gray gum, spotted gum, along with planted rainforest trees and plenty of animals. The gray gums bear scratches from possums and koalas. Low's house has been the site of visits from a variety of wild creatures. Once, in the night, he heard sounds in his bedroom and woke up the next morning to find a python on his nightstand, about a foot from his face. The noise had been the snake knocking his clock and other things off the table on its way to bed.

In his backyard, I hoped, was the bird I'd come for.

It's a scrubby little yard with no lawn to speak of. But in a corner is an enormous mound of leaf litter and soil built by one of the strangest birds on the planet—the brush turkey, famous for its bizarre form of parenting and for a mysterious talent that enables its unusual behavior.

Low is one of Australia's most renowned naturalists, but this bird did not single him out for his ecological bent. As anyone in his neighborhood can tell you, brush turkeys are everywhere, their numbers on the rise. Young males looking for new nesting grounds are finding it in the leafy suburbs of Brisbane and Sydney.

"This mound is poorly built, probably by a young male inexperienced in architecture," says Low. "Too much soil, not enough plant material." He kicks the mound a few times, dislodging a chunk. This will bring the turkey running, he says.

It does. We hear a rustling in the adjoining yard next door, and we duck up to the porch to watch the bird's return.

The brush turkey is big like American turkeys, with subdued colors except for an odd array of head and neck appendages, bright scarlet face and vivid yellow wattle, a pouch of loose skin circling the base of its neck, like it took a splash from a crew painting road lines. It uses the wattle in its courting display, to bellow out a deep *boom*. Despite its name and superficial physical similarities, the bird is no relative of American turkeys. It's the largest of the twenty-two species in the megapode family, called so on account of the size of their feet, which are used for digging.

This turkey is mostly silent, though we catch some soft clucks and grunts as it scratches around where Low kicked the mound. It seems to be scraping material *away* from the hole rather than filling it up, unbuilding rather than rebuilding. But two months earlier, the bird was hard at his construction work, raking up grass, plants, and leaf litter from a huge area of Low's yard and piling it up in his mound. "He was quite systematic," Low says. "He worked in precincts, cleaning one wedge of yard floor, and then another until it was all bare dirt up to the mound. It was both fascinating and awful because of the damage he was doing— ripping out grass, gouging holes in the ground."

Will Feeney, a bird researcher who lives outside of Brisbane, told me of a similar clash with a male brush turkey. He had just planted a garden—parsley, basil, rosemary, tomatoes, capsicums, passion fruit. That night, he heard some scratching in the yard. "Everyone had been warning us about the turkeys," he recalls, "but I said, it's okay, it'll be fine—didn't do anything last year, probably won't do anything this year." In the morning, he went out to his yard to discover that the whole garden had vanished into that turkey mound—kind of like Thanksgiving in reverse. "It destroyed probably $500 worth of garden," he says. "I've been having a bloody war with this turkey ever since, where I'd go down and dig a couple of buckets out of its mound to put back in my garden because it's great soil. And then it would come take it back."

A male brush turkey spends nearly three months building his mound, raking up two to four tons of leaf litter and soil to create a massive conical structure roughly the size of a car, three or four feet high and up to twenty-two feet in diameter. All of this effort he exerts for a single purpose: to create an incubator for his eggs.

In one respect all birds take the same approach to parenthood: They lay eggs incubated outside their bodies. Eggs have been described as the "perfect package," robust, self-sufficient, capable of doing one extraordinary thing—nourishing and protecting a developing chick. But here the uniformity ends.

Eggs themselves vary in size from the tiny .007-ounce Tic Tac eggs of hummingbirds, to the six-inch-long, 3-pound eggs of an ostrich, which require 120 pounds of pressure to smash. They vary in shape, as well, from the ellipse of a maleo, a large megapode, to the nearly perfect sphere of a brown hawk-owl and the teardrops of murres and shorebirds. Until recently, little was known about the reason for the wide variety of egg shapes or their function. Scientists thought that egg shape might have something to do with a bird's nest type—its size and form, cup or dome or simple scratch in the sand; or its location—tree limb or burrow or cliff edge, where it might roll off (a cliff-nesting bird, for instance, might have a conical egg that would roll in a tight circle); or the number of eggs in its clutch and how they might pack together to allow females to efficiently incubate.

That changed in 2017, when Mary Caswell Stoddard and her colleagues developed a computer program called EggxTractor to analyze the shapes of about 50,000 eggs representing 1,400 species. For each species, they also collected data on adult body mass, clutch number, diet, and nest location.* The analysis comparing the 1,400 species revealed that nest location has little to do with egg shape. Neither does nest size nor clutch size. Instead, three factors are strongly correlated with the shape of eggs: an adult bird's body mass, its evolutionary history, and, intriguingly, the ratio of a bird's wing length to its width—a proxy for flight ability. Why might egg shape be correlated with flight behavior? The team suspects that it has to do with adaptations for powered flight. They hypothesize that flight influences a bird's body plan. It's all about how aerodynamic and streamlined its skeleton and muscles must be, which influences what shape of egg will fit through its oviduct. Strong flyers like whooping cranes, shorebirds, and murres, with more streamlined bodies, tend to lay eggs that are more elongated and asymmetrical, while birds that fly little, such as limpkins and owls, have more symmetrical, spherical eggs. Owls that are better fliers have eggs that are slightly more elliptical. "But this doesn't mean that flight ability is the best predictor of egg shape for all birds," says Stoddard. "If you zoom in on a particular clade, such as shorebirds, you might find that other factors are better predictors. It's like saying that in vertebrates across the globe, camouflaged animals are generally brown or green, but when we look at vertebrates living in the Arctic, the rule doesn't hold—they're usually white to match the snow." Yet another example of the way birds resist sweeping generalizations.

---

* It's worth noting that Stoddard's study was made possible by a museum collection of eggs gathered by obsessive collectors in the late nineteenth to mid-twentieth century, who ferreted out eggs in concealed nests, blew out their contents, and assembled them in vast collections that were eventually given over to museums. While this has yielded wonderful insights like Stoddard's, the zealous collecting put many species at risk, and nowadays, egg collecting is illegal.

The nests in which those polymorphic eggs are laid are as various as the birds that build them, from the scant scrapings of an Arctic tern in sand or gravel, to the dark burrows of auklets and the elaborate woven nests of village weavers and southern masked weavers, hollow globes of green grass, each of which—it was recently discovered—carries a "signature weave" of the individual artist. The communal nests of sociable weavers may house as many as five hundred of the sparrow-size birds in little apartment-like chambers that stay cool in summer and warm in winter and sometimes harbor the nests of other species—finches, tits, lovebirds, barbets, and pygmy falcons—and can hold the weight of a sunbathing cheetah. The smallest nest on record is the bee hummingbird's, a wee inch in diameter, and the largest, a bald eagle nest in Saint Petersburg, Florida, that was ten feet wide and twenty feet deep and weighed close to three tons.

Scientists lately discovered that the open cup-shaped nest so familiar to us today probably evolved from the roofed variety in more than 60 percent of bird species. Domed nests were the ancestral form for all passerines, a group that includes all songbirds. Open nests likely evolved at least four different times. They're more vulnerable to predators and weather but simpler to construct, and the good visibility makes it easier for birds to spot and reject brood parasites.

Little grebes build a floating platform for their nests from sticks and aquatic vegetation, so it glides over water. Hornbills raise their chicks in tree cavities, which the male closes up with a barrier of mud, bird droppings and sticks, sealing the female inside with the young; only her beak can poke through to receive food brought by the male—mice, frogs, fruit. Several species of swifts in Brazil—sooty, white-chinned, white-collared—nest in seemingly impossible spots. Brazilian scientist Renata Biancalana, who studies these little-known birds with their sickle-shaped wings and high dancing flight, found that they build small cup-shaped nests on vertical rock walls behind waterfalls in damp cavities and crevices inaccessible to predators, wet, dripping, and subject to constant spray. Driving through the countryside in Kalimantan, Borneo, you can see huge three-story concrete structures with tiny windows—far larger than any house in the region. These are homes for the edible-nest swiftlet, a bird that makes its nest out of its own saliva hardened into a woven cup. The little white nests, about two inches across, are considered a delicacy in bird's nest soup, one of the world's most expensive foods, so they're "harvested" in buildings designed to resemble the enormous rock caves where the birds normally nest.

I once wandered through an exhibit of nests at a bird festival in Spain that featured surprising nesting materials. The great crested flycatcher and some other birds are known to weave shed snakeskins into their nests. Sparrows use dog hair and cigarette butts—the nicotine may drive away parasites—and red-breasted nuthatches ring the rims of their nests with toxic conifer pitch to trap and kill intruders. Flowerpeckers, small, active, brightly colored "chili pepper" birds, as they're called, build beautiful purse nests suspended from leafy twigs, made of leaves and grass fibers felted together with spiderwebs. Rufous horneros, large ovenbirds of South America, collect mud, straw, and manure and pack it onto a tree branch. When the sun bakes the mud, it creates a solid spherical structure with a hole in the center, like a clay oven. Montezuma oropendolas weave pendulous nests up to six feet long from vines and banana fibers. An African species of penduline tit also creates an elaborate pear-shaped nest from spiderwebs, nettles, grass, and animal hair. It's domed in shape but has two entrances—one false, which leads to a small false chamber, and a second, true entrance covered by a flap that seals with sticky spiderwebs.

There's no question that nest building in birds is remarkable behavior," says Sue Healy. "Perhaps just as remarkable is the prevailing view that it reflects nothing more than genes, that it's purely instinctive and governed by a simple set of rules."

Healy, a professor of zoology at the University of St. Andrews in Scotland, has been arguing for more than a decade that nest building in birds is anything but simple, requiring sophisticated cognitive abilities—and, in this way, should be considered more like toolmaking. After all, nest building involves creating a new structure from a large number of objects—sticks, mud, mosses, grasses, feathers, snakeskins, spider silk. It requires making informed decisions about location and choosing appropriate materials to shield young from the elements and protect against predators. And, often, it requires coordination and collaboration between male and female birds. Penduline tits, for instance, build together as a pair for two weeks—one working on the outer structure of the nest, the other on the lining. (It's the only sustained time they'll spend together, a fortnight's fling.) Healy also points to brain studies showing that when birds build nests, a number of neural circuits grow active, including pathways involved in motor learning, social behavior, and reward.

Healy is a champion of research in this area. "There are still big holes in our knowledge," she says. "We have nest descriptions for only around 75 percent of species. We know who builds nests, male or female or both in only 20 percent of species."

Healy wants to understand how a bird knows what nest to build and where to build it. "There's more and more evidence that birds build their nests flexibly, in response to weather conditions and the materials they have available in their environment," she says, "and that they use their own past experience in making decisions." That is, nest-building behavior is far from hardwired or fixed. At least in some cases, it may be learned.

The classic method for determining whether a behavior is purely instinctive or has a learned component is the so-called deprivation experiment—depriving an animal of an experience while it is young and then seeing whether it nonetheless performs the behavior flawlessly the first chance it gets. Healy points out that American robins and rose-breasted grosbeaks that are hand-raised can't construct a proper nest, suggesting that in the wild, experience is important to first-time nest builders in these species. More evidence: Birds that build elaborate nests learn and become better nest builders over time. The intricate weaving of weaverbirds—their skills at looping, winding, and interlocking grassy materials and creating bill-made knots to fasten and secure them—improves with practice. Moreover, those individual "signature" weaving patterns mastered by village and southern masked weaverbirds suggest the birds develop their own style over time, which is so distinct and consistent that scientists found they could identify the specific builder of ninety-six nests with about 80 percent accuracy.

Healy also notes that many birds modify their nest locations depending on their own experience. Successful breeding season? Stay in the same place. Unsuccessful? Move to a new spot for the next season. This is true for species ranging from northern flickers and brown thrashers to orange-breasted sunbirds and spotted antbirds. One recent study showed that if nest-plundering jays are in the area, a female hummingbird will situate her nest near or beneath a hawk nest. When hawks are present, jays forage higher up to avoid attack, creating a cone-shaped "enemy-free" nesting space beneath a hawk nest, where hummingbirds can raise their young in peace. Some birds also note the experience of other individuals in making nest-building decisions. When black-legged kittiwakes or piping plovers fail at a breeding attempt, they use information on the relative success or failure of breeding neighbors to decide where to build their next nest.

Birds adjust the way they build their nests in the face of threats from both predators and weather. Superb starlings, which build their nests inside thorny acacia trees, tend to choose trees defended by the most aggressive ant species in the region, "presumably because the ants help to protect the nest from predators," says ecologist Dustin Rubenstein. A house wren building its nest in a box with a large entrance hole that might be inviting to predators will construct a tall stick wall between the hole and the nest cup, probably to deter intruders. Sophie Edwards, one of Healy's students, found that zebra finches build different styles of nests at different temperatures. Birds building nests at temperatures of

64 degrees Fahrenheit add more material to their nests, making them 20 percent heavier than the nests of birds building at 86 degrees Fahrenheit. This may be hopeful news, suggesting that some birds show flexible nest-building behavior in response to ambient temperature, which may help them adapt to climate change.

The nests of piping plovers are nothing special—mere scrapes in the sand like the Arctic tern's—but their way of defending them is. If a cat steals up to a plover nest, a parent bird may feign a broken wing, dragging it half-open and to the side. The one time I saw this, I swore the bird was crippled or injured. But it's a ruse. The plover creates the illusion that it would be an easy target, capturing the cat's interest and luring it away from the nest before the bird flits off to safety.

There are other, more aggressive approaches to nest protection. It's common to hear stories of geese and swans hissing and rushing at intruders with flapping wings, of gulls pecking at—and pooping on—anyone approaching their nests. But nowhere is the nesting season quite as frightening as in Australia, where children on their way to school, postal workers, bicyclists, dog walkers, garbage collectors, people simply taking a stroll in the park live in terror of the extremely protective parenting of one fierce bird: the Australian magpie.

Beware entering the range of a magpie with chicks. Most Australians have learned this the hard way. Some 85 percent have been attacked at some point in their lives, says Darryl Jones, a professor of behavioral ecology at Griffith University in Brisbane, known as the "Magpie Guy." His namesake is no relation to the European magpie, a member of the corvid family, but rather belongs to an endemic Australasian family that includes currawongs and butcherbirds. Of these, only the magpies are swoopers.

In greater Brisbane alone, 800 to 1,200 attacks are documented each year—and those are just the ones that are reported. I was there in late August and early September of 2017, just as the nesting season began,

and the stories I heard were chilling. The birds swoop in, often from behind, and hammer your head, neck, and face with their powerful beak
and scratch with their claws. Injuries are common, usually minor cuts
and scratches, but sometimes more serious harm occurs—broken limbs
or damaged eyes. "There are thousands of people injured every year,"
says Jones. "People have terrible accidents on bikes. And every year eyes
are lost. So it's a genuine issue."

Magpie Alert, a social website that tracks magpie encounters, invites
people to report attacks and calculates the statistics. The year I was there,
the site tallied 3,642 attacks and 591 injuries. The comments section of
the site is telling. On the day I'm writing this, "MQ" reports: "This
same magpie last year attacked me and left me bleeding. Today it attacked my daughter 3 times and left her badly bleeding and terrified."

The swoopers are almost always male. And only certain males are hy-
peraggressive in this way, says Jones—around 10 percent. "If it were
higher than that, I don't think Australia would be habitable." Males
swoop only when chicks are in the nest, a period of about six weeks, and
then the swooping stops abruptly.

It's not clear why the birds consider humans such a threat. "Cats and
dogs and snakes and whatever, that makes complete sense," says Jones.
"But as far as I know, most of the people who have been attacked have
never climbed a tree and eaten a baby magpie. So that's a huge question.
I think they're just hypersensitive."

The magpies seem to specialize. Around 50 percent of them are pedestrian magpies—they only swoop pedestrians. Cyclist magpies swoop
only cyclists. And then there are magpies that specialize in hammering
posties—the postal workers on little motorbikes that go whizzing up the
street, postbox to postbox. In his research, Jones found that magpies that
attack cyclists will attack any cyclist. And those that attack posties will
attack any postie.

"We think the cyclist and postie magpies are generalizing," says
Jones. He believes it has something to do with speed. "They'll be coming after you, and if you stop and get off the bike and walk it, they sort

of look around as if to say, 'Where is that fast thing that I was just chasing?'"

But the pedestrian magpies attack only certain people. "That's the other big thing we discovered," says Jones. "These birds recognize individuals. They live in a little territory and they never leave it, so they know every human that lives around the place. They see kids growing up, and they remember them. These birds are very, very smart. They're watching us and interpreting our behavior." The magpies live on average for twenty years and can remember up to thirty human faces for about that long, says Jones, "so if you anger a magpie once, you're going to get attacked again and again."

None of this stopped Australians from voting the magpie their favorite bird, beloved for its early morning caroling, described as a "quardle ardle oodle" or "waddle giggle gargle," depending on whom you ask—though neither phrase captures the marvelous musicality of the bird's song. Magpies can be very tame and are often kept as pets. Jones had what he describes as a wonderful roguish childhood pet magpie named Gymmy. The magpie known as Penguin gained fame for its role in lifting a family in Sydney from despair. "Everybody loves them," says Jones. "Every second sporting team is called the Magpies."

The people of Vancouver are less enamored of the crows that defend their nests in similar fashion. Attacks during the nesting season there, April through July, range from a single dive-bombing bird that draws blood to whole roving gangs of birds that pursue individuals for blocks. These birds, too, are expert at recognizing human faces and hold grudges against those they perceive as a threat to their nest. Jim O'Leary developed the Canadian version of Magpie Alert, called CrowTrax, which maps the 1,500 or so crow attacks across the city each nesting season. He says that it's the element of surprise that makes the attacks so unpleasant. "It's not like when you see a dog growling, where you can prepare," he told a reporter at the *Vancouver Courier*. "Crows generally attack you from behind. And it's very offending."

Swooping magpies and crows are just protecting their nests, says

Jones. "The swooping is a signal, saying keep away. The simplest solution? Avoid that spot. Never go back there. You'll be fine."

From fierce protection to complete neglect, the spectrum of parenting strategies in the bird world boggles the mind. Some birds minimize parental care, using geothermal heat to incubate their eggs or handing over care to other species. Others rely on biparental care, female-only care, male-only care, and cooperative breeding (birds banding together to raise their young). There is also a cornucopia of anomalies and special cases.

Biparental care by a pair-bonded male and female is the most common practice, occurring in more than 80 percent of species. Some years ago, when Andrew Cockburn estimated the prevalence of each mode of parenting in birds, he found that two forms of care were far more common than scientists had thought: female-only care (which occurs in more than 770 species, around 8 percent) and cooperative breeding (occurring in more than 850 species, or 9 percent)—four times higher than previous estimates. In Australia, cooperative breeding occurs in close to 1 in 5 species. The early lowball estimate is another example of blinkering by European naturalists—and also, the low density of scientists who study the vast and distinctive biota of Australia. Based on recent DNA studies, some scientists suggest that cooperative breeding, like dome nests and female song, may have been the norm among the ancestors of many bird families.

Single parenting by the female is most common in birds that feed largely on fruit and nectar from tropical trees, which occur in abundance. Male care for these species is of limited value, says Cockburn, "allowing females to choose freely among males for good genes rather than for direct benefits from the male such as a high-quality territory or paternal provisioning." Many of these are bird species that lek, such as hummingbirds, manakins, bowerbirds, and lyrebirds. The female picks a mate from the crowd and then goes off to raise her brood on her own.

It's harder to find a common pattern among the ninety species in which dads raise the young alone. They range from emus and cassowaries, which incubate their young—the cassowary for more than fifty days without leaving the nest or even standing up—to the comb-crested jacana, which is known to bundle his chicks up under his wings if they're in danger, ferrying them to safety with only their dangling legs visible.

As for the anomalies and special cases: Those penduline tits that collaborate to make their teardrop nests seem to have an oddball parenting system. Either the male parent or the female parent may desert the nest during the egg-laying phase, leaving the remaining partner in sole charge of the young. In a whopping third of cases, both parents desert the nest, abdicating all parental care.

Then there are the infamous eclectus parrots, with mothers that kill their male offspring.

Because of their gorgeous Christmas tree plumage, eclectus parrots are popular as pets. But no one had conducted field studies of the species until Robert Heinsohn began his research in 1997. Not surprising, really. The birds nest in hollows up to one hundred feet high in the canopy of remote rainforests in New Guinea and on the tip of Cape York Peninsula, where Heinsohn studied them. They probably didn't find their way to Australia by flying there, Heinsohn says. The parrots are not strong flyers, and it's a seventy-mile strait between New Guinea and Cape York. He thinks they probably found their way to Australia more than ten thousand years ago, when sea levels were lower and land bridges connected the land masses.

It took Heinsohn eight years of grueling, hair-raising fieldwork to unravel the dual mystery of their strange coloring and behavior.

The first nests he found were in hollows in the "Smugglers Fig," a magnificent old green fig tree that still had rusted metal spikes poking from its trunk put there by smugglers so they could climb up the fig to collect the parrot chicks for the pet market. Among Heinsohn's early discoveries was that females almost never leave their nest hollows, even when their chicks have matured, and even after they've fledged. Good hollows—big enough to hold a family of big parrots, high up in a tall

emergent rainforest tree towering above its neighbors (the better to protect against pythons and other predators) and illuminated by bright
sunlight—are so few and far between that females stay at their nest tree
for up to eleven months to defend them from usurpers. "Not only are
nest trees rare, but a lot of the hollows in them flood too regularly to be
useful for raising young," says Heinsohn. "Even large chicks can drown
when the hollows fill with water during torrential rain. So females have
evolved this sort of stay-at-home thing and do everything they can—
including fighting to the death—to hang on to their hollows."

The only way a female can manage this, says Heinsohn, is to have
males come and feed her. Three, four, five, up to seven males at a time
will comb the forest for fruit and when they return to the hollow, lock
bills with her and regurgitate the pulp and seeds. When Heinsohn looked
at the DNA of the males attending a single female, he found they were
not related—nor were they all fathers of her chicks. "Males have to compete very vigorously to get access to a female with a good hollow," he
says. "And they end up sharing her. She mates with a lot of them to make
them all think they're dads. So she has all of these males going out and
finding food to feed her, and that allows her to stay put and protect that
all-important resource."

Males, meanwhile, also visit up to five other females to try for sex.

"It's a mating system unlike that of any other parrot," says Heinsohn—
without the social monogamy and shared parental duties seen in most
parrots—and it all comes down to the scarcity of nest hollows.

This is also at the root of their exceptional coloring. Among most
hollow-nesting birds, such as parrots, the plumage of the sexes is often
similar: Because the female is concealed in the nest hollow, she doesn't
have to blend into her surroundings, so sexual selection for bright colors
or fancy plumage has carried both sexes in the same direction. But in the
case of the eclectus parrot, the roles of male and female have diverged
so much—the one being purely defensive and the other purely about
foraging—that they have completely different selection pressures work-

ing on them. This is why they break all bird coloration rules. "Their extreme reversed coloration is not linked with reversed sex roles but instead with stiff competition for rare nest hollows," says Heinsohn. "It's all about scarcity."

Females evolved a brilliant vermilion plumage, visible for miles against the leafy green branches, that acts as a gaudy beacon to other females, saying, "This hollow is occupied!" Males, on the other hand, evolved plumage of a rainforest green on their backs, which offers camouflage protection from aerial predators while they forage for themselves and their mates.

Heinsohn discovered the males' plumage was also laced with ultraviolet coloration, an important factor during courtship. "They are green and camouflaged when they want to be. But the moment they step into bright sunlight—around the nesting holes, for instance—the ultraviolet takes off, and they become very bright in the eyes of other parrots," including the females they're courting. "It's a clever compromise between camouflage from their enemies when they need it and showiness to their own kind."

The bizarre behavior sometimes observed in mothers—killing their sons just after they hatch—also arises from the shortage of hollows. Not all nesting hollows are created equal, says Heinsohn. There are dry ones, which are great for breeding, and there are poor ones with a tendency to flood in heavy rain, drowning eggs and chicks. Female eclectus chicks generally fledge up to a week earlier than males, so mothers stand a better chance of successfully reproducing by pouring all of their maternal efforts into female chicks. If a female nesting in a flood-prone hollow hatches a male and female chick, she will get rid of the male chick in order to speed the development of his female sibling. And because the chicks come out of the egg with their gender color differences, mothers can decide their fate within hours of hatching. It happens rarely enough that it doesn't affect the balance between the sexes in the population, says Heinsohn. "Provided you don't do it too often, you can get away with it."

The eclectus may take the prize for most unusual mothering. But anecdotes of unexpected parenting configurations among birds abound. Two male bald eagles and a female were observed tending a nest on the Mississippi River. A webcam in Nevada delivered news in 2018 of a pair of co-parenting female great horned owls—the first time this behavior has ever been observed in the species. That year, a photographer on Lake Bemidji in Minnesota spotted a female common merganser with seventy-six ducklings, which could not all have been her own. Rather, she was likely an older female experienced at raising young, and other females left their ducklings to her care in a kind of day care system called a crèche.

At the wildlife sanctuary just down the road from me, there's a female crow named Juno that came to the center with a damaged wing about fifteen years ago. She can't fly but she acts as a surrogate mother for the baby crows that are brought to the sanctuary each spring, literally taking them under her wing, feeding them, training them. "She teaches them how to be in the world so they can take care of themselves when they're released," one staff member told me.

Doesn't the tenderness of this bird just turn you inside out?

Birds also foster the young of other species. Wander through ornithologist Marilyn Muszalski Shy's classic review on incidents of interspecific parenting, and you'll find stories of Arctic loons raising a brood of spectacled eiders, a great horned owl taking charge of three red-tailed hawks, a song sparrow and a cardinal raising a brood of both species. I was especially warmed by reports of the vigorous and persistent feeding of one bird by another of a different species: a male eastern bluebird feeding house wren nestlings, even fighting the wren parents to feed the young birds; a house wren, in turn, feeding nestling common flickers while his mate incubated the eggs on his own nest—and continuing to feed them even after his own eggs had hatched. Another house wren filled a birdhouse with nesting material but then could not find a mate, so he fed three black-headed grosbeaks with caterpillars and other foods until they fledged. As if that wasn't enough, he then fed a family of nest-ling house sparrows. When a brood of eastern kingbirds were orphaned after an electrical storm, an eastern wood pewee fed the foundlings for ten days until they fledged. Likewise, a pair of starlings that lost their nest numerous times to a flood of rainwater in the drain spout they had chosen as their nesting site finally gave up their own breeding efforts and tended to some nestling American robins.

More examples have popped up lately. In a Michigan park, a pair of sandhill cranes were observed raising a gosling along with their own colt. And in Colorado, photographers caught a pygmy nuthatch feeding mountain bluebird nestlings and also doing a bit of cleaning for them, removing fecal sacs from the nest. Two scientists in northern Delaware reported the first known case of a wood thrush offering food and sanitation care to a nest of veeries, even though the parents were present. The researchers caught all the activity on video, and when they analyzed the footage, they discovered that the thrush was outdoing the parents, tidy-ing up more vigorously than either veery, disposing of twenty-six fecal sacs to the male veery's six, and delivering food to the four nestlings seventy-eight times, compared with thirty-three deliveries by the female veery and fifteen by the male. It also fed the runt of the clutch signifi-cantly more than its siblings.

Why would birds foster the young of other species? Shy mentions Richard Dawkins's suggestion in *The Selfish Gene* that adoption may represent a misfiring of parental instincts and "confers no reproductive benefits on the foster parents but instead wastes time and energy that could be invested in their own kin." Or it could offer practice in what Dawkins calls the "art of childrearing." Shy suggests there might be proximate causes—that the nest contained offspring from both species, or the helper's nest was destroyed or situated close to the "recipient" nest, or the calls of hungry young, no matter what the species, may serve as an irresistible summons. This seems to me the most likely hypothesis, that instances of interspecific nurturing are simply testaments to the power-ful pull of parenting.

I'm intrigued by the two poles of bird parenthood—group parenting and no parenting. In nearly 10 percent of species, from white-winged choughs to acorn woodpeckers, multiple adults band together to dote on their young until they're fully independent, sometimes for many months, even years. On the other hand, a small number of species, around 1 percent of birds, save themselves the trouble of parenting altogether, outsourcing the job to other birds. An even smaller group lets the earth do their incubating work and leaves their young to completely fend for themselves.

The brush turkey is one of these.

"These birds are remarkable in every way, but especially in the way they breed," says Darryl Jones. "They helped persuade John Gould that Australia abounds in anomalies." Known as the "Turkey Whisperer" in addition to the "Magpie Guy," Jones has spent the past three decades unraveling the reproductive lives of these birds. When he first came to Brisbane in the 1990s to do his PhD on brush turkeys, he had to travel to the rainforests of Papua New Guinea to find them. Now, they've moved into populated areas everywhere across southeast Queensland and are thriving as urban birds, building their enormous mounds in parks, gardens, backyards, even driveways.

Whether in the rainforest or the suburbs, the building of an incubation mound is a Herculean effort that takes a toll on a male brush turkey's body. He may lose up to 20 percent of his body weight while building. And the turkeys rarely use the same mound two years in a row, so each breeding season, their efforts begin anew. Not surprisingly, a male will defend his mound ferociously and won't leave it unattended for long, even to eat. Tim Low and I needn't have worried.

"These birds are the most spectacular engineers of any animal because individuals can produce such a monumental change in an ecosystem," says Jones. "In the rainforest, the organic matter on the humus layer is the engine that drives the nutrition for the forest," he explains. "Brush turkeys scrape up all of the organic matter from a big circle with a one-hundred-meter radius, all the seeds, all the fruit, everything goes into the mound and is compacted in that one place, so they're actually manipulating the *structure* of the forest."

Wherever it's located, in rainforest or suburban backyard, the mound acts just like an enormous compost heap. The heat is generated by fermenting vegetation, the metabolic activity and decomposition of a vast and diverse community of fungi, bacteria, and tiny invertebrates that thrive in the moist environment inside the mound. "So they're really cleverly harnessing this natural process," says Jones. "But it's a machine; it needs to be fed stuff to be maintained at a constant temperature."

Which is where the male brush turkey's secret talent comes in: Every morning, he excavates a narrow hole at the base of the mound and thrusts his head into the warm material to take its temperature. He can detect the slightest shifts in temperature and manipulate it by adding or removing fresh, damp vegetable matter, holding the temperature of his incubating eggs very close to the optimum of 91 degrees Fahrenheit.

How he acts as such a sensitive thermometer is not clear. Various body parts have been proposed as the sensor: feet, bare skin on the head, neck sac. But Jones says it's most likely in the palate or tongue. The birds take beak-fulls of material while working on the mound. "You can see them tasting it," he says. "They take a bite and then"—he clucks a few times—"manipulate it around their mouth, pushing it up against the palate." Depending on the result, they open up the mound and add material to adjust the temperature.

The turkey also pays close attention to the vicissitudes of weather and shifts the shape of his mound accordingly. If it rains heavily, the mound may saturate and suddenly drop in temperature, so he piles the top high to aid the runoff of rain and, after the rain stops, opens up the mound so it will dry out.

Malleefowl, the desert-dwelling megapode cousins of the brush turkey, are the masters of sophisticated temperature manipulation. Named for the low-growing bushy eucalyptus trees they use as litter for their mounds, malleefowl employ both microbes and the sun to heat their mounds. In spring, it's fermentation. In summer, it's solar heat, but the bird must prevent overheating in the hot months by piling the mound high with sand. In autumn, when solar heat diminishes, a malleefowl keeps its mound warm by spreading the sand out during the day to warm it and then piling it back up at night. Each time the bird opens its mound, it has to move close to a ton of material.

"Really, these birds have no business building mounds in such an arid zone," says Jones. "It's hard enough in the rainforest." They probably moved into the central parts of Australia millennia ago, when these areas were wetter. In the outback, mounds of sand sixty-five-feet wide are thought to be the work of extinct giant malleefowl. "It was a good idea at the time, but then Australia just progressively dried out," says Jones. "Now they're the absolute kings of temperature manipulation, but they're also endangered."

Male megapodes build all this real estate to entice females. But it's the females that determine when conditions are right to lay the eggs in the mounds and—most important—who gets to fertilize them.

To return to the topic of love and mate choice for a moment: There's no pair-bonding in brush turkeys. Both males and females have sex with lots of partners. And males have the equipment to prove it, says Jones, a "large and dramatic 'sperm delivery' device—a strange double-headed phallus so conspicuous that it allows sexing within days of hatching."

Jones claims that female brush turkeys make the most informed mate choice of any bird. "And I mean it," he says. "*Any* bird." Here is his thinking: "Because they don't have any parental care demands, no incubation requirements, all they need to worry about is producing eggs," he says. And they produce a lot of them, on average around twelve, but as many as twenty, one egg every five to seven days, "depending on how much nutrition is around the place," says Jones. These are big eggs, weighing more than half a pound—which, in total, amounts to about three times a female's body weight. Most birds choose one mate per season, Jones's argument goes. They mate once, and that mating fertilizes the whole clutch. "But female brush turkeys make a fertility choice for every single one of those twelve to twenty eggs. Each time they lay one, they make a very informed decision about which male will get to fertilize it."

A female shops around, paying frequent visits to assess the various mounds around her and their attendant males, as many as a dozen or more each day. She's looking in part for a male with good genes, so he needs to show off his colors and other physiological attributes, such as his ability to boom using his bright yellow wattle. (The wattle expands, fills up with air "so his whole throat bulges out massively," says Jones, and then he forces the air through his nostrils and makes a low-frequency booming sound, just like a sage grouse.) But mainly, she's scoping out a good, stable incubator. The structures themselves tend to look alike, so she may use competence of the mound owner as a measure, favoring males who defend their mounds and tend to it assiduously. "He must be diligent and resourceful, so the incubator works," explains Jones. "And he must be present at his mound and able to defend it."

Some males just don't measure up. "They build giant mounds that look to be all perfect but never get any females coming to visit, and we don't know why," says Jones. "We think they build these giant mounds because they aren't drawing females, and they've got nothing else to do. 'Well, no females today, so what's to do? I'll just add a little more to my already oversize mound.'"

Once a female has settled on a male and his mound, she'll fight other females viciously for access to it, pecking, biting, and delivering wing blows. Then she'll mate. The turkeys copulate on top of the mound. It's hardly a romantic encounter—more like mallards than cranes. The male walks up the female's back, grasping her by the skin of her neck lest she try to escape. "Males are nearly always aggressive," says Jones. They rush the females and peck at them, sometimes lightly, sometimes so violently that the female must shield herself with a raised wing—and even so, loses feathers.

To lay her eggs, she slowly excavates a deep conical hole in the center of the mound—with no help from the male, who frequently disrupts her work and harasses her by pecking at her head. After thirty to forty-five minutes of digging, she scoops up a bill full of warm substrate to check the temperature in the mound—she, too, has this mysterious skill. It takes her only a few minutes to lay her eggs before the male chases her off the mound and fills the hole. Away she goes, never to interact with her eggs or young again. You can hardly blame her for not hanging around given the graces of her mate.

Meanwhile, the male continues tending and protecting his mound. He'll fight off anything that threatens his eggs, including six-foot-long monitor lizards called goannas, which sometimes raid the nests, digging up the mounds and pinching the eggs. "The turkeys kick things into their face," says Jones. "They've got very strong legs." Jones has spent his career digging up mounds himself to measure eggs and chicks. "Usually the males are standing a meter away pounding me with sticks and rocks."

Over the next forty-five days or so, while Dad is futzing around with leaf litter and probing for heat with his bill and kicking away giant goannas, the embryos in those eggs buried deep inside the mound are growing in an environment of extreme humidity and high gas pressures. When they're ready to hatch, the chicks use their backs and feet to bust through their shell.

What happens next is miraculous: The chicks hatch into a dark, damp, suffocating world, entombed—quite literally—under tons of stuff. There's little oxygen and plenty of toxic carbon dioxide. "They start life at the bottom of a meter of dirt and sticks and rocks," says Jones. Each chick lies on its back and, using its feet, scrapes at the material above, compressing the dislodged stuff beneath its back. It spends as long as two and a half days doing this, digging upward little by little, struggling through the dense dirt and litter of the mound, pausing frequently to recover.

When the baby bird finally breaks into the light of day, it pops out as the most advanced chick of any bird species, instantly able to fend for itself, to run, feed, and fly. "Brush turkey chicks are so super precocial they almost require their own terminology," says Jones. They need to be.

You would think after all this that the chicks' father might be waiting with a mouthful of food. But no. He's actually a threat. "If a chick is unlucky enough to meet Dad on the way out," says Jones, "he has no idea what the chick is. It's just some horrible thing in his mound. And—I've seen this many times—he just boots it off into the bush."

The chicks survive this paternal hurling. What they don't survive—and this is the absolute cost of having no parental care, says Jones—are the cats and dogs and foxes and goannas and everything else around the place intent on eating them. "There's no parent to tell them what a predator looks like, what's an appropriate behavior if a cat comes up, how to hide, what to do." Some 97 percent of brush turkey chicks don't make it through the first week.

It seems an incredibly unlikely and inefficient way to reproduce. But apparently it works.

Fifty years ago in the Brisbane area, "brush-turkeys were little more than memories," writes Tim Low, "shadowy inhabitants of the early rainforest lingering on as an occasional wayfarer on forest fringes." Only the odd brush turkey passed through. Now the birds own virtually every suburb around the city. Their rise has been described as a turkey tsunami.

"In the past twenty years, the population has increased 700 percent," says Jones. "I don't have to go to the jungles of remote New Britain to study them; I can walk out my back door. I would never have predicted that these birds would become a successful urban species. I mean they're *rainforest* birds. It's hard enough in the forest; in the city, there are cats and dogs and swimming pools and cars—so many more pressures—and yet I think it must be the sheer weight of numbers. If every female's laying twenty eggs, then you only need two to maintain population. If more than two survive, then the population grows, and that must be what's happening."

At least in brush turkeys and other megapodes, one parent puts some energy into the whole parenting enterprise. With brood parasites, both parents forgo all duties, laying their eggs in the nests of other species and leaving to them all the labor of raising their young. They're known as the deadbeats of the bird breeding world. But is that fair?

Chapter Thirteen

THE WORLD'S BEST BIRDWATCHERS

In the woodlands of northern Mozambique, a Yao honey-hunter moves through the bush, uttering a loud call, a *brrrrr* trill, followed by a short *hm!* At first, nothing. But then out of the trees a wild bird about the size of a starling appears. Dark backed, with a pale breast, it approaches the hunter. What ensues is a kind of collaborative treasure hunt. The bird, a greater honeyguide, responds to the hunter's call with its own distinctive chattering call that differs from its territorial song, a sharp *tirr-tirr-tirr*, as it flies from tree to tree, leading the hunter through the forest. It flits from perch to perch, emitting its persistent call. The hunter follows. This pattern of leading and following is repeated until the bird reaches the treasure: a nest of bees ensconced high in a broad-leaved tree or tucked into a rock crevice or termite mound. Once close to the nest, the honeyguide perches nearby and lets out an "indication call," softer in tone, with longer pauses between notes.

Not just any human call will elicit this behavior. Among the Yao, it's that trill-grunt call, *brrrrr-hm!* For Hadza-speaking people in Tanzania, it's a melodious whistle. In recent experiments, evolutionary biologist

Claire Spottiswoode and her colleagues played the trill-grunt call of the Yao honey-hunter, an unrelated Yao call, or the song of a ring-necked dove and then tracked the birds' responses. In two-thirds of the trials with the trill-grunt calls, the birds responded by guiding the honey-hunters, leading them straight to a bee nest in more than 80 percent of the forays.

In this way, birds and humans communicate. "The honey-hunters use these special calls to signal honeyguides they're eager to follow, and honeyguides use this information to choose partners who are likely to be good collaborators," says Spottiswoode. It can work the other way, as well, with a bird initially summoning human partners with its special loud chattering come-hither call.

The bird literally points the way to hidden bee nests; from its calls and its flight and perching pattern, the honey-hunters construe the direction and distance to the nest. After the hunters harvest the honey (subduing the bees with smoke), they leave the energy-rich honeycomb wax for the birds to feast on.

The greater honeyguide earned its Latin name *Indicator indicator* from this unique—and probably ancient—cooperative relationship with African honey-hunters, first confirmed by Kenyan ecologist Hussein Isack in the 1980s. Both parties benefit, fair and square. The birds, which specialize on a diet of beeswax, can't get at much wax without the hunters' help subduing the bees and opening their nests, and the hunters often need the help of honeyguides to quickly and efficiently find the nests—an important source of calories for them.

As far as we know, no other wild animal collaborates with humans in such a direct way. "Of all the relationships between people and wild animals," writes Elizabeth Pennisi in the journal *Science*, "few are more heartwarming."

Heartwarming from a human perspective, perhaps. But honeyguides are Jekyll and Hydes. Less touching is their penchant for siblicide, for murdering their stepsiblings in the dark nests of their adoptive parents.

Greater honeyguides are brood parasites, those good-for-nothing moms and dads at the zero-parenting end of the bird spectrum. Females lay their eggs in the nests of other species and trick the unwitting host parents into raising the parasitic nestlings, often at the expense of their own young.

Brood parasitism is extremely rare in the vertebrate world. It occurs in a few kinds of fish and in some species of ants, bees, wasps, beetles, and butterflies, but no reptiles or mammals. In the bird world, just 1 percent of species, around one hundred, are so-called obligate parasites, dependent on this way of reproducing. The strategy is risky, but for some birds, it seems to work. It has evolved independently at least seven times in different bird lineages: in ducks, honeyguides, finches, and cowbirds, and three times in different lines of cuckoos.

In recent years, scientists have found that this unusual breeding system is generating some of the most intriguing behaviors in the animal kingdom: sophisticated trickery, keen powers of pattern detection, ingenious covert systems for sharing information, and complex strategies for decision-making. All thanks to a tit-for-tat war between parasite and host.

Greater honeyguides often target as their host the little bee-eater, a beautiful cinnamon bird about the size of a sparrow, with a bright green backside and yellow throat. It nests in narrow tunnels dug into the roof of much larger holes excavated by aardvarks. The female honeyguide incubates her egg inside her body for an extra day before dumping it in the bee-eater nest. This means that her chick will hatch ahead of the host eggs, the better to ready itself for what comes next: murdering its nest mates. The weapon is a pair of needle-sharp hooks at the tips of the chick's beak, which it uses to stab its foster siblings one by one as they hatch. (One intrepid zoologist tested the honeyguide's biting power on his own flesh and reported that he had his tongue punctured by the upper hook.)

Scientists have known for decades that honeyguides kill their nest mates. But the truly dastardly nature of the deed had never been witnessed until Claire Spottiswoode and her team filmed the killings in situ, under natural conditions, using cameras positioned in the dark underground nests of the bee-eaters.

The dim infrared footage of the video recordings feels downright Hitchcockian, like watching an über-creepy bird version of *Psycho*. Blind and in total darkness, the honeyguide chick feels around for its newly hatched nest mates, snapping at the air haphazardly until it strikes flesh. Then it repeatedly grasps its victims and bites and shakes them until they die from hemorrhaging beneath the skin. Just a few minutes of active biting time can kill a bee-eater chick, but sometimes death is a slow, brutal process that takes up to seven hours.

Seeing the film, you can't understand why the host parent bee-eaters nearby don't hop to the aid of their bullied and badgered—their *pierced*—young. But it's dark down there, and bee-eaters don't have much in the way of night vision. The footage even shows host parents attempting to feed a honeyguide chick that is busily savaging their own bee-eater young. Such highly effective killing behavior ensures that the honeyguide chick alone survives to monopolize the nest and the parental care of its unwitting hosts.

Honeyguides are not alone in this heinous behavior. Lots of brood parasite chicks kill their host's young soon after they hatch. The striped cuckoos of Latin America use the honeyguide's technique of stabbing their nest mates with their needlelike bill hooks. A common cuckoo chick adapts the "push from a height" strategy seen in Hitchcock's *Rear Window*, hoisting a host egg or even a newly hatched chick onto a small indentation in the middle of its back and then, with its legs braced against the sides, tipping it up and out of the nest—all with its eyes closed. The chick looks like a malevolent little wrestler hoisting a small child up and heaving it over the window ledge. That a tiny, blind, naked hatchling can be so murderous seems astounding.

Parasitic chicks have other tricks up their sleeves. To dupe their host parents into bringing them more food, chicks of the Horsfield's hawk-cuckoo, an Asian species, flash bright yellow patches beneath their wings when their host parents feed them. The wing patches look just like the gaping mouths of extra chicks and also brilliantly reflect ultraviolet light, signals that together make for an irresistible stimulus to host parents to feed the cuckoo nestling more than the standard chick rations. If all goes according to plan, the parasitic chick, gleaning the usual feedings, plus some, grows and grows, sometimes to monstrous size compared with its host parent. The sight of a European wren perching on the back of a giant common cuckoo chick ten times its size in order to get food in its mouth, or a petite white-plumed honeyeater struggling to satisfy the appetites of the massive pallid cuckoo chick overflowing its nest is among the strangest in nature and tears at the heart.

The British naturalist Gilbert White called the cuckoo way of reproducing "a monstrous outrage on maternal affection" and "a violence on instinct." How do these birds get away with such an outlandish departure from normal breeding behavior? Why are the host species so easily duped, blithely incubating foreign eggs and feeding chicks sometimes grossly—almost comically—different from their own? Why don't they eject the impostors and rescue their own offspring from such untimely and violent ends?

G rowing up, I viewed the brood parasites in my own neighborhood, brown-headed cowbirds, as truly bad characters. Prowlers and cheats, they stalked the hard-built nests of the "good" birds I loved— vireos, warblers, finches, and flycatchers—sneaking through the woods and thickets searching for an unguarded nest in which to dump their eggs, flushing the little host birds from their brooding by making short, noisy flights that mimicked the flights of predatory hawks.

When a female cowbird finds a nest and the timing isn't quite right for her to lay, she may destroy the egg clutches and broods, forcing the hosts to start over so they're in synchrony with her schedule. The chicks that hatch from her eggs don't normally toss out the host young that share their nest, but they do compete vigorously with them side by side. Many of the smaller host chicks don't stand a chance against the cowbird chick and end up dying of starvation.

Ornithologist Arthur Cleveland Bent called the cowbird a "shiftless vagabond and imposter." They even look shifty, with their brown overhead masks. Somehow, they can sucker more than 170 different bird species—ranging in size from kinglets to meadowlarks—into forsaking their own offspring to raise little cowbirds.

How does a baby brown-headed cowbird know it's a cowbird and not a kinglet or a meadowlark? When nestlings are "cross-fostered," put in the nests of other species, most species imprint on the foster parent and learn its behaviors, songs, even mate choice decisions. "Yet somehow brood parasites avoid this mis-imprinting," says Matthew McKim Louder, who studied brown-headed cowbirds at the University of Illinois at Urbana-Champaign. "They develop almost a sense of self," he says. For decades, it was thought this was a simple matter of instinct. But in 2019, Louder and his colleagues published findings that brown-headed cowbird mothers use a special "chatter call" that gives nestlings a key bit of information: self-knowledge. "The female brown-headed cowbirds produce the chatter call in different social contexts," says Louder. "The call acts as an acoustic password cue for her offspring, guiding their choices about who they learn from, so they learn the songs of their species and identify appropriate mates." Hearing the password actually changes the brains of the young birds, transforming the auditory region of the brain into a state ready for learning the songs of their own kind. In this way, a mother cowbird tells young cowbirds who they are and what they should become—a brood parasite rather than a member of their host species.

For their hosts, the brown-headed cowbird is nothing short of a nightmare. Researchers recently found that a cowbird female will closely observe the nests she lays in, and if the hosts eject the foreign eggs, she may wreak havoc on the nest and savage the eggs by puncturing them—possibly to force the hosts to build a new nest, the better to parasitize it, a strategy called "farming." Or possibly in revenge, punishing the hosts by destroying their nest and the entire clutch within it. These mafia-like tactics may explain why some host birds don't oust the cowbird's parasitic eggs. A host may learn that if it doesn't play the game the cowbirds' way, the parasites may come back and badly mess with their nests. And because the cowbird is such a highly prolific egg producer, laying up to forty eggs in a season (compared with a typical host species such as a red-eyed vireo, which tops out at ten), the opportunities abound for this kind of retaliation.

The cowbirds are good at what they do—so good that they appear to be contributing to the demise of dozens of already troubled North American songbird species on the brink of extinction from habitat degradation, including the endangered least Bell's vireo and the Kirtland's warbler.

But these are the dark facets of a much more complex, intriguing—even beautiful—tale. Brood parasites may be nature's greatest reproductive cheats, but they are hardly lazy or shiftless. Females in particular show cunning and courage. And their way of reproducing has led to a fierce and fascinating coevolutionary arms race that is pushing both parasites and hosts to extremes of clever adaptation and behavior. The race is changing the very nature of communication within a species—spurring the development of unique visual cues and vocal passwords—and also, possibly, communication *among* species, generating something like a universal bird language. It may even be driving the evolution of new species. And all so rapidly that it seems to be happening right before our eyes.

In a scrubland northwest of Brisbane, Will Feeney is exploring this epic race. Feeney, an expert on the interactions of cuckoos and their hosts, supervises a project on the nature of brood parasites and their effect on the birds they parasitize for the Environmental Futures Research Institute of Griffith University. He and his team are here for the complex community of parasitic cuckoos and their hosts that breed in these parts—birds tightly bound in a life-and-death race that shapes their bodies, eggs, and offspring, their behavior, brains, and more.

It's a different sort of race than scientists thought. Until lately, much of what we knew about brood parasites we knew only from studies in Europe, which has two species of cuckoos, and in North America, which has one. "In those two areas, you get three out of a hundred species in the world," says Feeney, "and yet something like 90 percent of the research on brood parasites has been done there. In Australia, we have ten species of brood parasites—seven of them at this site." From Feeney's research thus far, it appears that all seven species parasitize more than one host, and most hosts get hit by several different species of cuckoos. "When it comes to brood parasites, this place may be more reflective of the world at large," he says, "multiple cuckoos posing a threat to multiple different hosts with very interesting interactions between them. So we can ask, what's really happening in these systems? Is it really all about this neat coevolutionary arms race between two species, a host and a parasite, as we've assumed? Or is it more like Darwin's concept of an entangled bank, multiplied many times over, where many species of cuckoos parasitizing multiple species of hosts in an overlapping way drive not only the evolution of individual species but of entire host communities?"

This is one of the puzzles Feeney is trying to understand at his field site. "It offers the chance to blow open new questions about how birds relate to one another," he says, "how information transfers not just between individuals of the same species but between species, which is exciting. It's also so complex, which makes it kind of a gamble. But we thought, bugger, let's give it a crack."

Feeney's site lies in a swath of parkland in the shadow of Mount Nebo and Mount Glorious and bordered on one edge by Lake Samsonvale. It is hardly a pristine landscape, largely old pasture and farmland, once cultivated, now fallow and overgrown with shrubs—perfect habitat for the little ground-nesting birds Feeney studies, fairy-wrens and scrubwrens and the cuckoos that plague them.

Lake Samsonvale is man-made. In the 1960s, Brisbane bought up the land, forcing people off it, and built a big dam to create a reservoir, a source of water for the city. When rain is scarce, the water level in the lake drops so low you can see the old railroad tracks running through it. Fences abound, along with the cattle they keep in. Invasive plants such as lantana, a large flowering shrub native to Central and South America and considered one of the worst weeds in Australia, form dense, impenetrable thickets. Still, there are striking stands of hoop pines here, a species that has survived from the time of the dinosaurs, and patches of rainforest with majestic old fig trees. When the trees are fruiting, it's an eBird hot spot, with more than 240 species on record, an unusual mix of bush and lake-margin species. When things get dry, it's even a refuge for deep inland species, such as the pallid cuckoo. Feeney says that people flying through Brisbane will pop out here during a layover to pick up a pale-headed rosella or a pair of pardalotes—spotted or striated—and maybe a brilliant flock of scarlet myzomelas, Australia's smallest honey-eater. The occasional domestic bird can also be found wandering about, such as the helmeted guinea fowl someone abandoned at an old cemetery. (Feeney points out its uncanny resemblance to Jar Jar Binks, the clumsy outcast in *Star Wars*. The likeness is so striking you have to wonder if the character wasn't modeled on the bird.)

On one side of the arms race are the fairy-wrens, scrubwrens, and other small parasitized species. On the other, seven kinds of cuckoos, from the little bronze, a bird about the size of a sparrow, to the world's largest brood parasite, the two-foot-long channel-billed cuckoo, which usually lays its eggs in the nests of big birds such as pied currawongs, Australian magpies, crows, and ravens.

At the moment, Feeney is focused on the fan-tailed cuckoo, a slim blue-gray bird with a pale orange throat that parasitizes several species of fairy-wrens, scrubwrens, and thornbills, particularly the brown thornbill. He's also looking at bronze-cuckoos, including the Horsfield's bronze, a small nomadic cuckoo that moves into a breeding territory for just a few weeks and lays a string of single eggs in ten or twelve different host nests—mostly those of fairy-wrens. The nestling cuckoo usually hatches a day or two before the fairy-wren young and immediately sets about evicting all host eggs and nestlings. The wrens thus lose all their own young and then invest close to fifty days rearing nothing but the parasitic cuckoo chick.

Fairy-wrens are impressive little birds by any measure, feisty, full of nervous energy, fierce defenders of their nests. The term *wren* has been used for centuries in the Old World to describe one of the smallest of birds there, the Eurasian wren. The fairy-wrens of Australia were thus tagged by settlers nostalgic for their homeland and understandably smitten with these similar spunky but completely unrelated little birds in their new country. The "fairy" part popped up sometime in the twentieth century and stuck, despite ornithologist Ian Rowley's dismissal of the term as "too prissy for the average Australian to use in everyday speech." Cunning and bold, constantly in motion, they remind me of the sassy little Carolina wrens that live in my region, except that the fairy-wrens have a long, expressive tail cocked like a mockingbird's and, unlike North American and European wrens, are supremely colorful. Male superb and splendid fairy-wrens are splashed with brilliant blue caps and backs. The variegated wears an azure crown and ear coverts. The red-backed carries a dash of crimson and the purple-crowned, a regal mauve coronet. In the words of fairy-wren researcher Andrew Katsis, "They look like they were dressed by Cirque du Soleil."

As we walk, the calls and songs of the various fairy-wren species bubble up from the eucalyptus and acacia scrubland: the pipping of superbs, which escalates into a brisk boisterous rippling reel; the metallic trill of variegateds, like a tinny typewriter; the weak high-pitched reeling song of a red-backed. The fairy-wrens are all facultative cooperative breeders (meaning they don't necessarily breed this way all the time but rather in response to circumstances), with pairs often supported by helpers, who assist in defending the nest and feeding the young. In a typical breeding season here, from August to February, a female fairy-wren will build as many as eight dome-shaped nests and lay clutches of three or four eggs in each, which she incubates alone. This may sound excessive, as if the world should be overrun by little harlequins. But predators will get two-thirds of her young. And if she's not careful, brood parasites will get the rest.

Feeney is tracking both sides of the arms race and the profound effects they have on the birds' bodies, brains, and behavior.

We walk single file along a path that wanders through dry weeds and shrubs, looking for small orange flags that mark the nests found by the ten "nest-finders" on Feeney's research team. A black-shouldered kite passes overhead. Then a whistling kite. It's late August, end of the Austral winter, on the cusp of spring. A warm, rainy winter, good for insect irruptions, launched an early breeding season here. Last year, the Horsfield's bronze-cuckoos laid their first egg on November 11, toward the end of Australia's spring. This year, the laying started so early in the season that Feeney barely had time to get his research team in place. Some of the wrens and cuckoos have already fledged—the few that survived the onslaught of predators, snakes, antechinus, lace monitors, goshawks, kites, currawongs, little falcons, whipbirds, and kookaburras.

It's a tough road to successful reproduction here for both cuckoo and host.

"I think around 10 to 15 percent of nests in the UK fail because of predation," Feeney tells me. "Here, it's closer to 70 or 80 percent. I've had years where the first fifty or so nests fledged no young because of predators. I once saw an eastern whipbird just tearing into a nest, all beak and fury."

As we walk, I notice that the American members of Feeney's team are wearing high rubber gaiters. When I ask why, they exchange glances. "Snakes." None of the Australian team members seem fazed by snakes, even by the deadly eastern brown snakes common around here. They thrive on the outskirts of nearly every town and city in Queensland, where they take credit for the most serious snakebites. If provoked, they rear up and bite savagely. Human fatalities are rare. But a recent news story reported that a man died less than an hour after being bitten by a brown snake in his suburban backyard while trying to protect his dog.

I look down at my jeans and sneakers.

"You'll be fine," Feeney says. I have to take him at his word. A Queenslander from birth, Feeney was a snake handler in his teens, charged with removing brown snakes from the crawl spaces beneath houses. He and his team have a nickname for the snakes: "danger noodles." Brown snakes are perceived as aggressive, he says, "but they're actually extremely shy and will get out of your way." He pauses. "And anyway, the brown has tiny fangs. Even if it were to bite your jeans, it wouldn't get through them."

The ground-nesting birds here are not so fortunate. Eastern brown snakes attack either through the nest entrance or through a hole they burrow in the back of the nests. Then they snatch whatever they find inside, eggs or chicks. On Feeney's website is a haiku written by fairy-wren researcher Lauren Smith:

Tiny fairywren
*Who is your father? *snake chomp**
Guess we'll never know

The nests are hard to see, especially those of scrubwrens, which tend to bury their nests deep in the shrubs and weeds. "We sit and wait, watch the birds, follow them," says James Kennerley, a PhD student under Feeney and Andrea Manica at the University of Cambridge. "Once, we saw a female scrubwren nipping milkweed fluff out of a pond and traced her back to a nest deep in a lantana shrub. We also develop an ear for the begging calls of chicks."

Feeney and Mike Webster of the Cornell Lab of Ornithology run the field site collaboratively. Together, the research teams find about seven hundred nests a year. These they monitor closely during the eight-month breeding season, checking them for eggs—host and cuckoo—and counting and photographing what they find. (Kennerley is creating a photographic library of the eggs.) They also pluck the cuckoo chicks from the nests, measure them, and weigh them by stuffing them upside down in a film canister and placing them on a scale. Then they band them and take blood samples. Their goal is to monitor in detail the breeding successes of the species locked in a tight contest for survival, to understand who's winning, who's losing, and how they're playing the game.

Even with its orange flag, the nest of a red-backed fairy-wren is tough to spot—almost completely invisible, tucked into dense vegetation, thick grasses and weeds. Feeney slides two fingers into a small hole in the tiny, delicate dome-like mass of dried grasses, leaves, and sticks and gently pulls out three eggs, all pinkish white, covered with fine red-brown speckling. Two belong to the host; one, slightly larger, belongs to the Horsfield's bronze-cuckoo.

Eggs have traditionally been seen as the main battleground of brood parasite and host in their evolutionary arms race. The story goes something like this: A brood parasite targets a new host species, duping its victim into accepting its egg and raising its offspring. Eventually, the host evolves the ability to detect the foreign eggs and reject them. The parasite then evolves eggs that mimic the color and pattern of the host's eggs so closely that they escape detection. In turn, the host has two defenses: change the look of its own eggs, or heighten its ability to distinguish them from those of the interloper. And so it goes. It's a high-stakes game of evolutionary ping-pong. For every defense evolved by their hosts, cuckoos retaliate by evolving ever trickier means of fooling them. Each move is met by a countermove.

This process of reciprocal adaptation gives rise to some astonishing results.

For one thing, cuckoos around the world have evolved superb egg forgeries. Their hosts, in turn, have evolved such exceptional powers of discrimination that they can spot a cuckoo egg that's almost completely

identical to their own, with just the tiniest discrepancies. It's a sophisticated learning process—less about spotting the odd egg in their nest and
more about learning in detail the look of their own eggs.

Some host species have evolved an additional defense: eggs with a
unique pattern of spots and squiggles that varies from female to female.
Mary Caswell Stoddard and her colleagues recently explored how several
species of British songbirds that are targets of the parasitic common
cuckoo might fight back against cuckoo egg mimicry with specially pat-
terned eggs. "If hosts are having a hard time fighting cuckoos," she says,
"one way to make the job easier is to evolve a complicated and really
recognizable pattern that allows you to say, 'That egg is mine.'"

Do host birds lay eggs stamped with their own distinctive patterns?

"To answer this question, I knew we were going to need what amounts
to facial recognition software for eggs," says Stoddard. The differences in
pattern and color in these host egg types are so subtle that they often
can't be detected by human observation. "You can't pick up the difference by eye—you really need a computer to analyze them," she says. So
Stoddard teamed up with a computer scientist, Chris Town of the University of Cambridge, and created NaturePatternMatch, software that
identifies features on an egg that are likely important to a bird. It regis-
ters these features the way a map might register important natural
features—lakes, mountaintops, groups of trees—only the egg features
are more like blobs or squiggles. "One might look like Mickey Mouse,
and the program is able to say, I've seen that Mickey Mouse blob before,
even if the image is rotated and the blob is not quite in the same place or
the same size," explains Stoddard. "It's a really robust recognition algo-
rithm that analyzes patterns in a way that roughly resembles how birds
and other animals may process visual information."

The program yielded just what Stoddard and her colleagues had
hoped: clear evidence that several host species have added to their eggshells individualized signatures, "just as a bank adds special watermarks
to its dollar bills," she says. The variation from individual to individual makes the egg patterns difficult to forge. Some of the strongest egg

signatures occur in those species for which cuckoos have evolved very close mimicry for both egg color and pattern, such as the brambling, a member of the finch family.

These "egg races" are moving at an extraordinary pace. Claire Spottiswoode and her colleague Martin Stevens of the University of Exeter found that eggs laid by two African bird species, the parasitic cuckoo finch and its host, the tawny-flanked prinia, have changed in color and spot pattern over a period of just forty years—a mere blink of the eye in evolutionary time—and have been closely tracking each other's evolution.

"At the heart of the egg war is a visual question that's a matter of life or death," says Stoddard. "A cuckoo's egg mimicry has to be spot on. Otherwise, its egg is going to be detected and rejected. Likewise, for the host, if it fails to recognize and reject a cuckoo egg, it's game over. So you have to have excellent egg mimicry on the part of the cuckoo and excellent egg recognition on the part of the host."

And that's where it was thought to end. In the common cuckoo—and every other brood parasite, it was assumed—it all came down to eggs.

"It's true that a lot of the work around the world focuses on eggs," says Will Feeney. If a common cuckoo or a cuckoo finch gets its egg into the nest of a brambling or an African prinia, and the egg passes muster, the brood parasites are all set. The hosts of common cuckoos almost never reject cuckoo chicks—even though they look nothing like host young. Even a few weeks after hatching, when a cuckoo chick is more developed, the host birds are strangely incapable of identifying the little monsters that have hijacked their nest, and will care for "Rosemary's baby" until it fledges.

There's a theoretical model suggesting why this is so. Bird parents "imprint" on the chicks that hatch from the eggs in their very first clutch, and after that, reject any chick that's different. Host parents that are unlucky enough to be parasitized in their first clutch will imprint on the cuckoo chick and rebuff their own young forever after. The high cost of mistakenly imprinting on a cuckoo nestling should prevent host birds from evolving the ability to reject chicks. This theory, born in Europe and North America, was thought to apply to all brood parasites.

"But things are a bit different here in Australia," says Feeney.

James Kennerley reaches into the little domed nest of a scrubwren and finds two eggs. "Promising," he says. The eggs look similar but can easily be distinguished—one belongs to the scrubwren and one to a fan-tailed cuckoo. The cuckoo egg, like the scrubwren's, is white with reddish-brown speckling, but the cuckoo's speckles are sharp and well defined, whereas the scrubwren's speckling is just a mess of dots and blotches. Both have a concentrated ring of speckles at the base. Hors-field's bronze-cuckoos also lay eggs that do a pretty good job of mimicry, but in general, Australian cuckoo eggs are poor mimics of host eggs compared with their overseas counterparts. Shining bronze and little bronze-cuckoos don't rely on mimicry at all. They just try to hide their eggs, coating them with a thick layer of dark pigment. To the host bird's eye, the pigment makes the eggs indistinguishable in color and lumi-nance from the lining of its own nest, effectively making them invisible.

Fairy-wrens for the most part can't reject or desert foreign eggs by their appearance. Especially if the cuckoo female has timed her egg lay-ing to coincide with that of her fairy-wren host. If she can't reliably spot a cuckoo egg in the dim confines of her nest, what's a female fairy-wren to do? Something it was believed no host bird ever did.

If a cuckoo parasitizes a host species, the host will evolve the ability to reject the foreign eggs. This was the accepted wisdom when Naomi Langmore of Australian National University started working on cuckoos in Australia in the early 2000s. But the more field studies she looked at, the more evidence she found against this. In fact, most Australian host species tended *not* to reject cuckoo eggs. "Which is completely counter to evolutionary theory," she says. "The arms race between them should exert powerful pressure on the host both to recognize foreign eggs and to reject them."

Were the host birds down under really incapable of distinguishing between their own eggs and a cuckoo's?

To find out, Langmore made a slew of little plastic eggs that were bright blue—different from the egg color of any host species in Australia—and placed them in the nests of a lot of different host species. She found that host birds that built cup-shaped nests could easily recognize and reject those plastic eggs. But birds that built dome-shaped nests, such as fairy-wrens, could not. This makes sense: Dome nests are dark places where it's tough to make visual distinctions.

Then Langmore made the sort of discovery that shocks scientists and makes research on this weird system of parenting so rewarding. When she looked more closely at birds that build these dome nests, primarily fairy-wrens and thornbills, she found to her amazement that although they couldn't reject cuckoo eggs, they had a very fine ability to recognize cuckoo chicks—and reject them. The little host birds would either seize the intruding chick and drag it out of the nest or abandon the nest altogether to start a new breeding attempt.

"This ran counter to everything we had learned about coevolution in brood parasitism," Langmore told me. "Hosts shouldn't evolve to reject cuckoo chicks because of the risk built into that imprinting mechanism. This, we thought, was the reason chick recognition had never evolved in hosts anywhere in the world."

Welcome to Australia, land of avian outlaws.

"Here, the arms race between cuckoo and host has escalated to where hosts can discriminate cuckoo nestlings," says Langmore. And here's the extra-cool thing, she told me. In Australia, not only do the hosts break the rules, so do the cuckoo chicks. "Because of this good recognition ability in hosts, cuckoo chicks have evolved wonderful mimicry of host young, either in appearance or in begging calls, which you don't get anywhere else." The bronze-cuckoos, which target fairy-wrens, thornbills, and gerygones, she says, "have evolved these fantastically matching chicks that mimic the size, the skin color, the down color, even the mouth color of the host species."

Horsfield's bronze-cuckoo chicks even *sound* like the host young, learning over time to match their begging calls. "This is pretty remarkable when you consider they evict the host chicks before they ever hear their calls," says Langmore. "They seem to attend to what sounds make the host parents most responsive and rapidly modify their begging calls over several days until they're virtually indistinguishable from the calls of the host young"—an extraordinary form of learning. The nestlings of screaming cowbirds, a South American species, also look and sound similar to the nestlings of their primary host, the greyish baywing.

Langmore suspects that this particular arm of the arms race—cuckoo chicks mimicking each of their hosts so closely—might even be driving speciation in the Australian cuckoos. "Each time one of the bronze-cuckoos adopts a new host, it has to evolve this different chick morphology and different begging call," she says, "and this might actually be driving them to diverge into different subspecies."

As for the evolution of counter-strategies in cuckoo hosts: Diane Colombelli-Négrel, a researcher at Flinders University in South Australia, and her colleagues discovered that fairy-wrens fight back against the cuckoos' begging-call mimicry in an ingenious way: They give their young a chance to learn a password while they're still in the egg.

Both superb and red-backed fairy-wren mothers produce special calls while they're incubating their eggs, singing them over and over, the same tune every few minutes. The calls contain a unique note that acts like a familial password, which the embryonic chicks commit to memory. Dads learn the passwords, too. After the chicks hatch and begin to beg for food, they reliably include that signature note in their begging calls, so their parents know, "Aha! *That* chick is mine."

To determine whether the begging call was learned or genetic, the scientists conducted a cross-fostering experiment, swapping clutches of eggs in twenty-two different nests. When the swapped eggs hatched, nestlings used the incubation call of their foster mother, not their biological mother.

It's not known why the cuckoo chicks, which are also exposed to the incubation song as eggs, don't learn the passwords. Perhaps they don't have sufficient time, only a couple of days' worth of training, or perhaps they don't possess the fairy-wren's embryonic learning abilities. In any case, the fairy-wren parents use the absence of the password in a cuckoo chick's begging call to single it out as an imposter.

Naomi Langmore and her colleagues found that superb fairy-wrens abandon close to 40 percent of Horsfield's bronze-cuckoo chicks within two to six days of hatching, and start building a new nest elsewhere. Females leave first. Males follow within a few hours or days, leaving the cuckoo nestling to die of starvation, its corpse eaten by meat ants within just a few hours.

The cost of mistakenly rejecting your own young is extremely high. To minimize the risk of making an error, female fairy-wrens seem to rely on a sophisticated decision-making process and a suite of complex rules, integrating clues from the nestling itself, from the wider environment, and from their own experience.

First, a single chick in the nest is a bad sign. Fairy-wrens have broods containing multiple chicks, so a lone nestling probably means the rest of the clutch has been evicted. So, too, an adult cuckoo at large in the neighborhood or near your own nest is ominous news. Fairy-wrens only reject cuckoo chicks when adult cuckoos are hanging about. Perhaps most important is an individual fairy-wren's personal experience. If she's had a chance to learn the look of her own chicks, she's less likely than a naïve female to reject her own nestling, even if it's alone in the nest.

The clues have to add up before a wren will take the risk of desertion.

Why have fairy-wrens evolved to recognize and abandon cuckoo young—and reed warblers haven't? "Fairy-wrens have 'lost' the arms race at the egg stage—they never evict cuckoo eggs," says Langmore, "and this has resulted in stronger selection for defenses at the chick stage."

Still, even in Australia, chick rejection is relatively rare, largely because it's so inefficient and so costly. For Australian hosts, the real defensive maneuvers against cuckoos occur early in the game, *before* the female cuckoo even lays her egg. It's far less ruinous to keep a cuckoo out of your nest to begin with. Feeney and his team are looking at the clever frontline defenses that hosts have evolved, including social learning and swapping information between species.

One August morning at dawn, Feeney leads me down a narrow path through stands of acacia, melaleuca, and eucalypts, toward a scrubwren nest buried in a bush. We hear the monotone *ee-ee-ee* of eastern yellow robins and the mournful descending trill of a fan-tailed cuckoo. He checks the nest. One egg, quite cold. "Hopefully it hasn't been abandoned."

If wrens see a cuckoo around their nest and discover an egg before their laying time, they may desert it or deconstruct the nest and move it to a spot nearby. Yellow warblers suspicious that a cowbird has laid in its nest will bury the cowbird egg, along with its own eggs, and build a new, superimposed nest on top. James Kennerley once observed a pair of wrens take the opposite tack: Apparently aware of the presence of a cuckoo and definitely aware of a new egg in their nest, "they disassembled the whole nest around the egg," Kennerley told me, "leaving the egg just sitting by itself in a shrub."

Feeney pops open a small dome-shaped camouflage blind around thirty feet from the nest, then unpacks a freeze-dried shining bronze-cuckoo, a tattered but recognizable model. He places the model in a protective cage of wire mesh and perches the cage a couple of meters from the nest. Then we retreat to the tent to wait and record what we hear.

When a scrubwren sees a cuckoo near its nest, it won't always physically attack the cuckoo but will unfailingly deliver a harsh chattering of scolding notes that may summon other birds. Superb fairy-wrens, on the other hand, tend to attack the cuckoos directly. This is the reason

Feeney's cuckoo models are all beat up. The birds go straight for the head with their sharp little beaks.

We hear the shrill notes of a mistletoe, Australia's only flowerpecker, and the *whip-chew* duet of a whipbird pair. But otherwise the bush is silent.

"This can be really boring," Feeney says. Superb fairy-wrens are attentive and vigilant near their nests, coming around to check on it every thirty minutes or so. They don't miss much that goes on in their neighborhood. Scrubwrens, on the other hand, may leave their eggs for three to four hours at a time. "You have to just wait until they bloody decide to turn up," says Feeney. This may be a defense against cuckoos, which use host activity to locate nests. "But it makes studying them hell," he says.

In one experiment on the social behavior of fairy-wrens, Feeney was tasked with observing juvenile males around the nest. "Juvenile males don't often go to the nest alone," he says, "and on one occasion, I had to sit for three days before I got a chance to run my experiment. They don't make much noise around the nest, really, so you have to pay attention. You just have to sit there like an idiot, sweating your head off, just watching the nest and doing nothing else. That experiment was hell." Plenty of other birds happened by during the study, including an emu that stuck its beak straight through the canvas of the blind, startling Feeney. "Bloody giant dinosaur bird! Straight out of the Jurassic."

After a half hour or so in our hide, Feeney gives up. "Bugger this. Let's go to the other nest."

We set up the blind in a little clearing in the woods near a nest where a female scrubwren has laid three eggs. Feeney again perches the cuckoo model nearby and retreats into the blind. But this time, he plays a recording of a scrubwren cuckoo alarm call. The reaction is instantaneous. The female scrubwren appears suddenly and unleashes a tirade of scolding notes. Other birds pop up in the low branches of surrounding shrubs. Feeney counts every bird that comes within fifteen feet of the cuckoo model. More scrubwrens, a Lewin's honeyeater, an eastern yellow robin, a gray fantail, a pair of superb fairy-wrens, and a couple of vociferous brown thornbills. Even a brush turkey strolls across the clearing. Relatively speaking, it's a modest showing. Feeney has had spectacular responses, where as many as thirty other birds show up to investigate the cause of the commotion in a kind of collective neighborhood outrage.

The superb fairy-wrens make the biggest ruckus, says Feeney. They come within a couple of meters of the cuckoo and "won't shut up." The mobbing calls they make toward cuckoos are not innate; they're learned. Feeney discovered this when he and Naomi Langmore conducted a version of the experiment Rob Magrath did with fairy-wrens and noisy miners. They looked at two different populations of fairy-wrens—one "naïve" population breeding in the botanic gardens of Canberra, where there are no cuckoos, and the other in nearby Campbell Park, a eucalypt woodland where cuckoos abound. They set up the cuckoo models just as Feeney had done with me. The naïve birds ignored the model, says Feeney. "They would check out the model and then just sit there or bounce past it and park themselves on their nest at the other side." Fairy-wrens in Campbell Park, on the other hand, aggressively attacked the models. "They'd go nuts toward the cuckoos," says Feeney. "My models are all missing their feathers from the back of their head because even if I put a cage over them, the fairy-wrens would squeeze under it to attack the cuckoo. And if I held up the cuckoo in my hand, they'd attack my hand. It's mental. You're getting this insane response from the one group and just ho-hum from the other."

Feeney and Langmore also found that young, cuckoo-naïve superb fairy-wrens could learn to recognize cuckoos as a threat by observing mobbing behavior of more experienced neighboring fairy-wrens. This is true, too, for reed warblers. The naïve birds at first completely ignored the cuckoo specimen, but after watching other fairy-wrens mob a cuckoo, they aggressively mobbed the model themselves. "This was the first direct evidence that naïve hosts can learn to identify brood parasites as enemies through social learning," says Feeney. The lesson sticks. Once a fairy-wren learns the identity of an enemy, it will respond the same way years later even if it has had no contact with cuckoos.

Feeney noticed that the mobbing calls the superbs were making toward the cuckoo had a very specific sound, "totally unmistakable," he
says. To find out whether the birds were really making cuckoo-specific
vocalizations, Feeney presented the fairy-wrens with models of all the
different threats they might encounter in the wild—snakes, currawongs,
hawks, and cuckoos—and recorded their responses.

"We found they absolutely lose it towards the cuckoos. To the snake
or the hawk, they made a few alarm calls during a five-minute trial
and then just shut up and watched the threat. To the cuckoo, they made
a huge number of calls, and they were different—more this whiny-
sounding call." It makes sense that birds would have a very particular,
very powerful call specifically for brood parasites, says Feeney. "When
you lose your nest to a predator, you can start building again tomorrow.
Whereas if you're parasitized, you're spending, what, six weeks of time
and effort raising a chick that is in no way contributing to your reproductive success. It's a different thing."

The call works to brings other birds in to help mob the threat. Mobbing can be very effective in thwarting cuckoos. Justin Welbergen and
Nick Davies of Cambridge University found that it lowers the chances of
parasitism fourfold in high-risk places where cuckoos hide and monitor
the comings and goings of hosts. A hard mobbing can damage a cuckoo
badly, causing feather loss, injury, even death. It can also discourage
cuckoos from laying by drawing hawks and other predator-enemies of
the cuckoos and by alerting neighboring hosts, which may join in the
mobbing and guard their nests more vociferously.

A few species of cuckoos have gotten wise to mobbing and have
evolved clever ways to manipulate it. Some mimic predatory hawks,
spooking host species into fleeing rather than mobbing. The common
hawk-cuckoo of Asia, popularly known as the "brainfever bird" because
of its loud, maddening call, looks just like a sparrowhawk called the shi-
kra and even mimics the raptor's flight style and the way it settles on a
perch. It's so convincing that when many birds see it, they raise the "bird
of prey" alarm not the cuckoo alarm. The common cuckoo also uses a

hawklike flight pattern when approaching a host's nest. And that's not the end of their craftiness. In 2017, Davies and his colleague Jenny York found that female common cuckoos utter the characteristic frightening laughter-like *kwik-kwik-kwik* or *kii-kiii-kii* call of a sparrowhawk to distract reed warblers from spotting the cuckoo eggs in their clutch.

Cowbirds have been known to employ the opposite—but equally manipulative—tactic of displaying an "invitation to preen" posture, which seems to lower aggression in their hosts.

Other brood parasites deter mobbing with a "wolf in sheep's clothing" strategy, mimicking the look of harmless species. In sub-Saharan Africa, one of those dastardly honeyguides, Wahlberg's, looks just like innocent grey flycatchers. Drongo cuckoos of southern Africa closely resemble the drongos that are their namesake, small birds that often steal food from mixed flocks but are tolerated by them because they act as predator sentinels. And cuckoo finches of Zambia have evolved plumage colors and patterns that look exactly like a benign bird common in the region, the southern red bishop. "The cuckoo finch is so similar to the innocent bishops, that the target of the trickery, the tawny-flanked prinia, can't tell them apart," says Feeney, "which suggests the cuckoo has evolved to be able to hang around prinia nests without arousing suspicion."

During the years Feeney was presenting cuckoo models to birds in Australia, he began to notice that cuckoo alarm calls from species to species sounded a lot alike. "I was asking myself, 'Was that thornbill just making a call that sounded like the fairy-wren's?'" he says. "It niggled at me, and I thought, 'Am I making that up?' Sometimes you're out there in the field and it feels like you're seeing or hearing something that isn't there." But later, when he was in Africa, observing the prinia's response to the cuckoo finch, it hit him. "I'm thinking, 'I could be crazy, but that prinia's call sounds just like a fairy-wren responding to a bronze-cuckoo.' And then, I was at this conference in Sweden grabbing a beer with a mate, David Wheatcroft of Uppsala University, who works in the

Himalayas, and I was describing these weird cuckoo calls and how simi-lar they are in different places, and this mate asks me, 'So what's it like?' And I tell him, and he says, 'Oh, does it sound like this?' And he pulls out his phone and plays a call for me. And I'm sitting there stunned, going, *Shut up*. And he says, 'These are my warblers responding to a cuckoo in the Himalayas.'"

Now Feeney is running trials and collecting cuckoo alarm calls from all over the world, including Africa, India, China, Indonesia, and Japan to see if birds from different countries respond to one another's cuckoo alarm calls.

"It would make sense from an evolutionary standpoint that host spe-cies in different parts of the world have converged on a common cuckoo alarm call," says Feeney. "Cuckoos are a unique kind of superweird threat—they're no danger to the adults, just to the young. So in thinking about the evolution of alarm calls, you ask, 'What are the pressures?' In response to a predator like a pied currawong, a bird has to walk that line between sounding the alarm and being spotted by the predator. 'I want my friends to know about this threat, but I don't want *it* to know about *me*.' Whereas with cuckoos, all bets are off, and you just want *everyone* to know about it and come in and kick the living hell out of that cuckoo. If that's the case, you'd expect selection to favor basically the loudest, most annoying sound you can come up with, with the highest chance of bring-ing in as many species as possible."

What does the call sound like? A sort of high-pitched whine. Imagine if a seahorse could neigh.

In Feeney's view, brood parasites may be driving convergence in an alarm call that effectively screams *cuckoo* in any bird language. "We know that information transfer about the risk posed by parasites is im-portant within a species," says Feeney, "but if parasites pose a threat to multiple species, it might make sense for information about the threat to also transfer *across* species—perhaps to facilitate a kind of behavioral herd immunity." He mentions a study published by a team of scien-tists in 2016 that analyzed the associations in sound and meaning in thousands of human languages. The team found a hundred or so archetypal words that are similar across continents and linguistic lineages—among them *star, leaf, knee, bone, tongue,* and *nose.* The distribution and history of these associations suggests that these archetypal words weren't inherited or borrowed but emerged independently.

Could the same thing be happening in birds? asks Feeney. "Could there be an international bird word for cuckoo?"

At one scrubwren nest, Feeney pulls out a cuckoo egg and shines his phone flashlight beneath it. The "candling" reveals the tiny squirming outline of a developing cuckoo chick. It's a common enough sight—but somehow also beautiful and miraculous. Especially when you consider what the female cuckoo had to do to get that egg where it is, developing in that feathered nest, beneath the warmth of a foster parent who, with any luck, may nurture it until it fledges.

"It's a really, really tough breeding strategy," says Naomi Langmore. "And it's funny that brood parasites have this popular public reputation for being cheats and being lazy. What they actually do is very difficult, very challenging, especially for the female."

Imagine being in her shoes.

"She really has her hands full just finding enough nests," says Langmore. Most evidence suggests that females do so by watching nest building, observing the movements of the adult hosts and following them, unnoticed, to the site—often extremely well hidden or cleverly placed.

Some female brood parasites are very choosy about the nests they target, selecting the best parents in a population—experienced birds with the highest likelihood of successfully raising their offspring. How do they manage that? Possibly by gauging a host pair's reproductive success the previous year—and remembering it. Brown-headed cowbird females are thought to size up the success of their offspring, returning to lay their eggs only in host nests that did well by their chicks and avoiding those that failed. Or, perhaps, by eavesdropping on the sexual signals of a mated pair, such as the quality of ornament, song, or display, how dominant they are over other birds in their group, how active they are in building their nest, how big and solid that nest grows—all honest signals of parental quality, and all information she can glean from a distance without rousing the ire of the neighborhood.

In Langmore's study population, each cuckoo pair parasitizes ten to fifteen fairy-wren nests, so females have to monitor the nest building of female fairy-wrens in more than a dozen spots. Some brood parasites have to contend with host species such as the yellow-rumped thornbill that build their nests near species that deter the parasites—beneath the large aeries of eagles or other predators or, like weaverbirds in Africa, close to the nests of stinging insects. Yellow warblers in North America nest near red-winged blackbirds, which are super aggressive toward the brown-headed cowbird.

For each nest, the cuckoo watches the female fairy-wren build and then, once she's finished, sneaks in when the parents are out and checks the nest to find out whether the fairy-wren has laid yet. Fairy-wrens lay in a short three-day window, so the female cuckoo must time her egg deposits just right. She has to lay her egg just when the fairy-wren is starting her own clutch, or her egg will be roundly rejected or even sewn into the lining of the nest. "She has to lay it so it's sufficiently in sync with the host's eggs," says James Kennerley. "If it hatches too soon, there's a good chance the host will abandon the nest. And, of course, if it's too late, then the chick probably isn't going to be able to compete against the host chicks for food."

The cuckoo checks the nests regularly to see how many eggs there are. When it comes time to lay, she will remove one egg and replace it with her own so that the wrens will be fooled into thinking their clutch is intact. Scrubwrens make the task trickier by laying only every other day, which may be a way to evade the cuckoos—if the hosts see that an additional egg is laid on one of the "in-between" days, that's a cue to abandon the nest.

In the actual laying of her egg, the cuckoo must be swift and stealthy.

She lands in a shrub a meter or two from the nest, then alights on the domed nest and enters through a hole at the front, her tail spread wide and still visible outside the hole. Quick as a wink, she lays, moves backward out of the nest, carrying the wren egg in her bill, and flies off.

Total time spent at the nest? About six seconds.

Then, no rest for the weary. She's off to search for nests in which to lay her eggs in subsequent days. She'll have to remember their precise location and the nesting stage of each and must avoid returning to lay eggs in nests she has already parasitized. Otherwise, she may get competition between her own offspring.

Finding the nests, watching them, carefully timing the laying of her eggs to overlap with egg laying in her hosts: all this while avoiding detection, which might set the whole neighborhood abuzz and rally mobbing. She has to do this ten or twelve times in a season, which means she has twelve pairs and twelve nests to keep track of. She tunes in to the activities of all of them, observing keenly, moving around and making mental notes.

Brood parasites may have smaller brains than their nonparasitic songbird cousins, "but they have fantastic spatial awareness," says Langmore. "It's perhaps a matter of allocating brain space differently. A songbird has all of this amazing neural circuitry devoted to song learning that cuckoos don't have. But brood parasites have a bigger hippocampus, a brain region heavily involved in learning and memory for spatial aspects of the environment." Studies on cowbirds show that females have more hippocampal volume than males, which makes sense, given that they are the nest searchers. Female cowbirds also outperform males on spatial memory tasks that resemble host nest visits in the wild—moving through space and remembering specific locations for a period of twenty-four hours.

Throughout my time in the field, I've found myself rooting for the little fairy-wrens and harshly judging the cuckoo's way of being. But the truth is, cuckoos have a much tougher go of it. Even if a female cuckoo manages to get all of her eggs into nests at the right time, most of them will be destroyed by predators—68 percent. And if the cuckoo chick hatches, then the fairy-wren's likely to reject it. "So they get very poor success," says Langmore, "and we're absolutely thrilled if we get three or four cuckoo chicks fledged in a year."

Spatial information, temporal information, social information, keen powers of observation—it's all going into a program the female cuckoo uses with precision, and against great odds, to perpetuate her species.

"What finely tuned creatures these are," says Langmore. "Really. Because of the bizarre evolutionary arms race they're engaged in, they've become the world's best birdwatchers."

Chapter Fourteen

A CHILDCARE COOPERATIVE
OF WITCHES AND
WATER BOILERS

Imagine you're flitting from branch to branch on the bird family tree, looking for a style of parenting diametrically opposed to the minimalist approach of the brush turkey or the fan-tailed cuckoo, a method of childrearing offering maximum attention, nurturing, devotion, and protection. You might not immediately settle on another twig belonging to the cuckoo tribe. After all, eagles, herons, emperor penguins, ospreys, and least terns are all known for spending excessive amounts of time and energy warming their eggs and caring for their young.

However, the greater ani is a special case. A big black bird with a long tail and a large bump, or keel, on its bill—which helped earn it the West Indian nickname of "witch bird"—the ani is a member of the cuckoo family, noisy and ungainly, with a "disjointed way of alighting in a bush as if thrown there," as naturalist Archie Carr once wrote. But it's renowned for its strange and wonderful system of collective childcare. And also for its highly unusual and endearing displays of group solidarity, right up there with the most spirited pep rallies.

Several times a day, the birds come together near their nest in a truly astonishing communal display unique to this species. "It's really like a giant football huddle," says Christina Riehl. The birds pile into a circular cluster, their bills all pointing toward the center, and then begin to chorus in unison—a long collective gurgling call like the hum of an outboard motor or bubbling water, which gave the greater ani its South American nickname, *hervidor*, or "water boiler." The collective call may go on for ten minutes or more.

An assistant professor of ecology and evolutionary biology at Princeton, Riehl has been studying this charming and disarming outlier of the cuckoo family for well over a decade. She focuses her intelligence and considerable courage on the task of unraveling three intriguing mysteries about the greater ani and what they might tell us about the lengths to which some birds will go to work together—and why.

In a video Riehl shows me, four big birds awkwardly jostle for space on a single basket-shaped nest of twigs on a limb overhanging a marshy river in Panama. In the nest is a heap of eggs from multiple sets of parents. The greater ani is a cooperative breeder. This way of parenting—where birds raise their young with help from siblings or offspring or other birds who forgo their own reproduction—was once considered exceedingly rare. Now we know that 9 percent of bird species breed cooperatively and share parental care, more than nine hundred species, including acorn woodpeckers, Florida scrub jays, western bluebirds, and many of Australia's little bands of fairy-wrens. Most cooperative breeders are "facultative," meaning they can breed successfully without helpers. Relatively few, like the ani, are "obligate"—they require helpers at the nest in order to successfully fledge young.

The behavior has baffled biologists for centuries. Think of it. Why would an individual give up its own breeding opportunity to act as a helper, assisting others to raise their young? Darwin worried that equivalent behavior in bees and other insects was possibly fatal to his theory of natural selection. The same William Hamilton who despaired over the coloring of the eclectus parrot is credited with rescuing Darwin's theory in the 1960s with his theory of inclusive fitness, also known as kin selection—

which suggests that helping your close relatives can be a good way to pass on your own genetic material because relatives share a percentage of your genes. The idea can be roughly understood by one biologist's remark (after scribbling on the back of an envelope) that he "was prepared to lay down his life for two brothers, four cousins or eight second cousins." By caring for kin, nonbreeding helpers are thought to gain "indirect" fitness: Even if they aren't passing on their genes directly through offspring, some of their genes will continue on (indirectly) in the family lineage.

Most cooperative breeders raise young in close-knit, family-based groups made up of genetic kin, with related helper birds from past broods assisting their parents in rearing younger nestlings.

Not so the greater anis. Group members of their parent cooperative are all genetically *unrelated*. Riehl learned this the hard way, by tracking nests in the wild, monitoring some forty to sixty communal nests, recording the fate of each over a dozen breeding seasons, taking blood samples from adult and young anis, and running DNA tests to identify the birds in each nest and determine the mother of each egg to reveal their genetic relationships.

Riehl is fascinated by strange reproductive and social systems in birds, especially cooperative breeding. When she started her PhD work a decade ago, "the assumption was that cooperative breeders were *always* related to one another," she says. "Cooperation evolved only among family members. It was all about kin selection: You help pass on your genes by helping relatives that breed."

As Riehl discovered, greater anis are rebels. Here is a bird that seems to prefer long-term living and nesting with strangers. This is the first mystery: Why would birds join up with unrelated birds to accomplish something as vital as raising their young?

Riehl says that she has loved birds since she was a child. "The thing I wanted to do more than anything was to go out with binoculars and identify as many birds as I could." After college, she worked in Kenya documenting helping behavior in the highly social turacos known as white-bellied go-away birds (their name comes from their distinctive *g'way!* call). When she went to Panama for the first time to study animal movement with her PhD adviser, she was drawn to greater anis, a very common bird, and surprised to find that no one had studied them in the wild. "I knew they weren't a good subject for the study of animal movement because they stay put," she says, "but they turned out to be a great subject for the study of weird social and breeding behavior."

Greater anis live along the forested banks of lakes, rivers, and streams in the moist tropical forests of the Amazon and its tributaries, from Argentina to Panama. Riehl studies them at the northern end of their range, on Barro Colorado Island in central Panama, where they abound in the marshy edges of the island.

"They are really gregarious birds," she says. In the nonbreeding season, they form huge communal roosts of a hundred birds or so. But in the rainy season, when it's time to raise their young, they coalesce into tight-knit groups of two to four unrelated, socially monogamous pairs, plus a couple of nonbreeding helpers. Together each group builds a shared nest in which all of the females collectively lay their eggs more or less synchronously. They all share parenting care, raising the mixed clutch of young until they fledge. The groups are extraordinarily stable, says Riehl, often staying together a decade or more. They don't migrate. Instead, she says, "they hang around together for long periods of time, so they have the opportunity to learn about one another and to experience parental care together."

Riehl shows me her photos of a nest lined with bright green leaves and holding the eggs of three females, all nonrelatives. In the nest is a pile of a dozen snowy white eggs with a lacing of blue. "Those eggs contain a cohort of unrelated siblings," she explains. "Three sibs from one pair, three sibs from another pair—unrelated nest mates all in the same nest."

The adult birds can't recognize their eggs or the offspring that hatch from them. Unlike the eggs of reed warblers or prinia, ani eggs carry no signature that would tell a bird, "This is my egg." "No one knows whose egg or whose nestling belongs to whom," says Riehl. And yet all of the adults share attentive parental care, warming the collective mound of eggs, taking turns looking out for predators, and feeding all the chicks with the big juicy grasshoppers, katydids, cicadas, spiders, required to stoke their rapid growth. Nestlings are born blind and naked, but by only five or six days of age are big enough to scramble out of the nest, swim, and climb. They'll leap from the nest into the water below, swim back to the base of the nest tree, and climb back up to the nest, hooking their bills over twigs as they go and using their wings to propel themselves upward. Adults watch over them vigilantly, taking turns as sentinels on the lookout for predators.

It's an extreme avian example of "We're all in this together," kind of like raising kids in a stable, well-run group house. Like any commune, it requires coordination and cooperation. How do these birds organize their efforts? Riehl wonders. And how do so many different individuals agree on what to do in the first place?

This is the second mystery. We know a lot about how animals in captivity learn, solve problems, and make choices together, she says, but very little about how they do it in the wild. "Honeybees, which dance to communicate information on potential hive sites—they're a great example of how many different individuals can arrive at a consensus to choose a new home."

How do anis do it?

The anis' most important group decision is where to build their nest. Safe nesting sites are few and far between, and they are key to breeding success. The birds prefer to nest over water, in marshy vegetation along lake and river edges, or in tree branches or curtains of lianas overhanging the water, places less vulnerable to snakes and monkeys—and, for similar reasons, seriously challenging to the field researcher.

Little was known about the greater ani before Riehl began her studies in 2006. David Davis, a biologist who tried to study the greater ani decades ago but gave up because of the difficulty, wrote, "Even though using a canoe at all times, it is frequently impossible to keep up with the birds or to follow them when they go into the thick brush or flooded lands along the river course." Riehl has the advantage of small flat-bottomed outboard motorboats that allow her to move slowly along the shoreline, following adults. In this way, she and her team have successfully monitored more than forty nests over the past decade or so, recording the fate of each egg and each nestling, collecting blood samples, and capturing the birds on video with nest cameras. Still there are huge logistical challenges and dangers, apart from the usual bees, wasps, mosquitoes, boa constrictors, and heatstroke. A forest of submerged trees, for instance.

Barro Colorado, the site of Riehl's field station, sits in the middle of Gatun Lake, an artificial reservoir formed in 1914 when the Chagres River was dammed to allow ships to transit in both directions. From the field site, you can see the giant container ships cruising back and forth on the wide waters. The island was once a hilltop surrounded by moist tropical forest, now isolated in the flooding created by the dam. The water around the island looks smooth and navigable, says Riehl, but just beneath its surface, there are hundreds of thousands of submerged hundred-year-old tree stumps. "You have to constantly weave in and out to avoid the stumps, and our boats have hit them and broken down more times than I can tell you."

The other major threat also lies just beneath the surface.

Crocodiles.

"We can't get in the water safely because there are crocodiles everywhere, huge crocodiles. So we do all of our work from the boats, but always keeping a wary eye out for them." This became a real challenge when Riehl and her team had to set up mist netting to catch adult anis along the shore. The researchers mounted nets on poles in shallow water parallel to the shore to capture birds as they flew over the water. "Mist nets are a pain," she says. "You have to get the tension just right between the poles and get the poles in just the right place in the water, and the motorboat is floating around, not staying still."

On one expedition, Riehl had just erected a mist net. An ani flew straight into it and was struggling to free itself. "We were using these little plastic kayaks to retrieve the birds," she says. "This is a very tricky business. You kayak up to the net, and you have to stand up in this tippy little boat to extract the bird from the webbing." On this occasion, she paddled to the net and began to stand up in her kayak, when an enormous crocodile suddenly exploded from the water. "It was like *Jaws*," says Riehl. "It took everything in its mouth, including the ani and the whole net, and just pulled it all down. And I'm in this little plastic kayak watching this and thinking, maybe this isn't such a good idea . . ."

After the US transferred the canal to Panama and ended its population control measures for crocodiles, their numbers rose. Trapping adult birds along the shoreline became increasingly dangerous. "We had to stop mist-netting adults for a while," says Riehl. "I was having nightmares about my field assistants going out in kayaks and getting flipped by a huge crocodile. But we've started up again; everyone is aware of the danger but very cautious."

While crocodiles may take an occasional adult ani, white-faced capuchin monkeys and snakes are the birds' real enemies. Riehl has found bird snakes in ani nests and rat snakes with recently swallowed eggs still visible in their bodies. Predation is very high in tropical birds in general, says Riehl, and in Panama, about 70 percent of ani nests are hit. Snakes are especially persistent, returning day after day until the nest is empty. Birds that belong to larger groups have a much better chance of escaping predation and reproducing successfully. They're even capable of killing small snakes.

This may be the answer to the first mystery: Unrelated adults join up for protection and to cooperatively mob nest predators. "The larger the group," says Riehl, "the more likely you are to fledge young from your nest."

But doing anything in a big group—especially something as complex as breeding, raising, and protecting chicks together—demands organized teamwork.

Think of the precise coordination it takes for a single pair of birds just to reproduce. Male and female join in a string of behaviors that prepare them for courting, mating, and building a nest. As early as the 1950s, studies on ringdoves showed just how tightly synchronized these behaviors must be. A female must hear and see her mate display in order to trigger her own urge for nest building and egg laying. Cues from the female in turn bring about changes in her mate that induce him to incubate. "It's this very precise interaction of behavior and display that pushes them both physiologically to the next stage of reproduction," says Riehl. Underlying these shifts are changes in reproductive hormones.

"Anis have a similar feedback process, but they have to do it with two or three different pairs." All the breeding birds in the group, males and females, four, six, eight birds together, have to settle on a nest site and synchronize their readiness for reproduction.

Then there's coordinating the labor of nest-tending duties: Who's going to sit on the eggs now? Who's going to feed the chicks now?

In birds that parent as pairs, bouts of incubation and food finding are tightly orchestrated. Many songbird parents, for instance, take turns brooding and feeding chicks. In the few short weeks after their eggs hatch, they each typically deliver around two thousand food items—insects, fruit, grubs—to fuel the growth of their young from tiny, naked, helpless hatchlings into fully feathered fliers some twenty times their original size.

For a long time, scientists thought that parent birds just provided a fixed amount of labor that was genetically determined, whether they were incubating eggs or feeding young. But in the 1980s and '90s, experiments on songbirds upended this idea. Parents took turns feeding young, visiting the nest at similar rates. But if you removed one member of the pair, the other would deliberately compensate, working harder to fill the gap. This suggests that each parent monitors the other's investment and decides how much care to provide. Sensitivity to a partner's workload varies, not just from species to species but from bird to bird. Blue-footed boobies are champions in this respect, equitably sharing duties. House sparrows have not quite hit this sweet spot of cooperation and egalitarianism, tending to ignore changes in their partner's work rate. (Unless they detect infidelity, in which case, they curb their own efforts. Studies show that male sparrows with mates prone to cheating provide less food for their young.) Great tits, on the other hand, are right on top of pair fairness, each parent flexibly adjusting to the behavior of its partner and matching its effort. Even among great tits, however, there's a "negotiation continuum," with some birds quick to respond to the behavior of other partners, and others, not so much.

It's not just songbirds that show this kind of subtle partner interplay in coordinating parenting. Little auk parents take turns going on long and short foraging trips. Kentish plovers, ground-nesting shorebirds that breed in the hot desert, coordinate their parenting efforts to protect the eggs—and themselves—from heat stress.

Multiply all this organization of efforts by two or three or four, and you see what the anis are facing. Riehl will tell you: Watching a group of these birds put it all together is fascinating. In most cooperatively breeding family-based bird groups, a single pair dominates, and all the other group members help raise the young of that pair. Greater anis are far more egalitarian. Everyone gets to reproduce. And they seem to have a team mentality in accomplishing every task. All group members pitch in to build the large bulky stick nest, pairs taking turns like in a relay, says Riehl. One couple will spend a few minutes to more than an hour—the male playing hod carrier to the female's bricklayer—until another couple arrives to take over the labor, and the first leaves abruptly, "as though by mutual agreement," says Riehl. Should any group member utter an alarm call, the whole team quickly gathers to collectively mob the threat.

Once the first egg appears, however, things may take a decidedly less cooperative turn. A female who hasn't started laying herself will unceremoniously remove any eggs she sees in the nest. Riehl shows me striking footage of the phenomenon. A female ani lands on the nest, cocks her head to inspect a single egg sitting there, then pulls it toward her with her bill, rolls it up onto the rim of the nest, and nudges it over the edge, *plunk*, into the water below. "It's a very clumsy process," notes Riehl. "I've seen females holding the egg between their legs and then, just after they've pushed the egg through, actually fall backwards off the nest."

This is the third mystery. Why would breeding females in these tight-knit and egalitarian cooperatives eject eggs laid by their fellow group members?

Other communal nesting birds have been caught committing the same act. Ostriches, which are ground layers, will roll the egg of other group members toward the edge of the nest—just to get them out of the way. Walt Koenig of Cornell University witnessed another joint nester, the acorn woodpecker, leaning out of its nest hole with an egg in its bill, about to drop it. The group nesting system of this woodpecker is among the most intricate of the vertebrate world, with family groups consisting of as many as seven related males and three related females, all laying in one nest, along with a slew of helpers. Like anis, female acorn woodpeckers that find an egg in their nest before they begin laying will destroy it, a practice that results in the demise of more than a third of all laid eggs. But, as Koenig and his colleague Ron Mumme discovered, the smashed eggs don't go to waste. The woodpecker group takes all the egg debris to a tree and feasts on it together, including the females who laid those eggs. Mumme calls this "communal egg sucking."

The female ani that lays an ejected egg doesn't resist the displacement. "She neither attempts to guard her eggs nor retaliates against the female that evicted it," says Riehl. Ani eggs are jumbo-size, almost 20 percent of a female's body mass, so her sacrifice is no small thing.

At least the arrangement is egalitarian. "Which unfortunate female lays first seems to be a toss-up," says Riehl. It doesn't have to do with her age or experience or how long she has been part of the group. Once a female lays her own egg, she stops ejecting the eggs of others, and once all the females have begun laying, the destruction ends altogether. Nature seems to throw a hormonal or physiological and mental switch, and all members of the group settle into caring for the whole communal clutch, coordinating all aspects of parenting, juggling duties with two or three pairs.

How to get everyone singing from the same song sheet?

Apparently, the same way that sports teams do—with frequent huddles. One bird gives a loud, distinctive call to summon its "teammates." It's a high-pitched heraldic cackling *kaa-kaa-kaa* that's given only in the context of these displays and can be made by any bird in the group. "It's a rallying or recruiting call," explains Riehl, "like 'Okay, guys, let's get it together, let's display, let's do this!'" All of the other group members fly in, sometimes from significant distances, and perch next to the calling bird, forming a circle, bills inward—except one bird, who sits outside on the edge to act as a sentinel, keeping a lookout for predators. The birds in the circle sometimes touch bills and may move in and out of the huddle, jockeying for position, climbing or hopping around one another, but always keeping their bills in the center. Then the heraldic bird transitions from the cackling call to that low-pitched mechanical gurgling, and everyone joins in.

Riehl observed more than 140 of these circular chorusing displays in more than 30 different groups and found that they occur multiple times in a day, with members flocking to the meetings and joining in the chorus. "It's clear they're participating in this loud, shared, synchronized signal together," says Riehl. "Is there meaning in the synchrony? Are some birds better at this than others? There's a whole lot of information in these vocal signals we don't understand yet."

Why the birds collect in these displays at all is also still largely a mystery. They're similar to the rallying choruses performed by cooperatively breeding green wood hoopoes of Africa, which function as territorial contests between neighboring groups, says Riehl. But greater anis rarely display in response to territorial encounters.

"We're trying to do some controlled experiments to figure out their exact function," she says. "My sense is that they have to do with coordinating the group. They display a lot when they're choosing their nest site and again right before they start to lay, and again after they've started laying eggs and the reproductive cycle has begun." Once laying has begun, they stop displaying almost completely.

Riehl suspects the displays have more than one function. "They probably reinforce social bonds between members," she says—not unlike regular team meetings. In other cooperatively breeding birds that live in family groups, the helpers are young birds that come and go, staying for a year and then leaving. By contrast, "greater anis stick together for a really long time," says Riehl. "We have groups that have been together since we started studying them in 2006." The displays may promote group cohesion or bonding after an encounter with a snake or a monkey, "like 'Okay, the predator was here but it left and we all survived, so everything's going to be fine.'" They may also have a deeper biological function ("I'm ready to lay eggs, are you?") and a deeper social function, facilitating group decision-making—about where to build a nest, for instance ("I like this spot for a nest, do you?"). Riehl's preliminary findings suggest that different pairs within the group often begin to build different nests within the group's territory, and that the displays play a role in determining which of these sites the group eventually chooses.

"But how do individual birds 'vote' in these collective forums?" Riehl wonders. "How do they overcome disagreements and conflicting opinions?"

Riehl recently discovered how the birds solve one aspect of the "many cooks" problem. When it comes to the all-important job of sitting on the big basket of eggs, they don't divide their labor equally. One male will do most of the incubation. And another serves as sole nocturnal incubator. "At night, it's always the same bird sitting on the nest," says Riehl. "This makes sense from the point of view of efficiency and safety. If every night you had to decide, 'Okay, who's going to warm and protect the eggs tonight?'—that's a really hard thing to coordinate with six or seven birds. It may come at a cost to that one male, but the benefit is that there's never any risk of the eggs being left exposed."

Riehl has also learned that there's great variation in how well groups work together. "Some groups are highly synchronized and on the same page," she says. "They make great decisions together and are really good at coordinating their parental care. Others really stink at it, and they can't figure out when to lay, can't synchronize their reproduction. Female A may lay an egg, which Female B removes. Then two days later, A lays another egg, and B pushes it out. Two days after that, A lays her third egg and B ejects it again. At some point female A will get fed up and say, 'I have laid five eggs and they don't seem to be here anymore so I'm going to leave!' This kind of asynchrony can lead to the dissolution of the whole group, which spells disaster for all. They toss out their eggs, their young get eaten, and the members end up abandoning their group."

Why do some groups cooperate and cohere, and others fall apart? "We don't know yet," says Riehl, "but it's exciting that anis are giving us the opportunity to explore how they work together in these complicated social groups and what makes them successful."

Living in a complex social world can have a powerful effect on cognition in a bird. Scientists have found that social living in animals actually changes the brain, enhancing the neural mechanisms underlying cognition—the synthesis and release of neuromodulators, the formation and strengthening of synapses between neurons, the birth of new neurons. Group living demands a lot of cognitive skills, including learning, memory, perhaps even the ability to take another bird's perspective. It may require recognizing members and remembering your past social interactions with them, anticipating their behaviors, even grasping the relationships between other members of your group.

Birds that live in big groups generally have developed complex signaling repertoires, whether vocal or visual, and may learn to modify their vocal signals to encode new and more detailed information. Their problem-solving abilities are enhanced, too. We know from studies of honeybees and fish that pooling information from more individuals improves performance in solving problems. Birds that work together to tackle challenges may have a leg up on those that try to do so alone. This may point to why many Australian birds are so extreme in their behavior and so intelligent. They often live as residents in complex societies, establish lifelong bonds with their group mates, and have long breeding seasons, so their interactions are collectively more complex than among birds in the Northern Hemisphere, which are often migratory and live together in pairs and only for a relatively brief breeding season.

New research on another cooperative breeder, Australian magpies of western Australia, suggests that when it comes to developing intelligence, the more birds in a group the merrier. Alex Thornton of the University of Exeter and his colleagues at the University of Western Australia, Ben Ashton and Amanda Ridley, tested cognition in wild magpies living in groups of different sizes, from three to twelve birds. The researchers found that birds living in larger groups were better at solving cognitive puzzles than those living in smaller groups. The team tested the magpies' ability to link a particular color with the presence of food and to remember where food was hidden, and also the birds' capacity for self-control—how quickly they could stop themselves from pecking at a transparent barrier and instead go around it to get at food—which is a reliable measure of intelligence. Birds that had grown up in larger groups learned and remembered their lessons faster and were better able to control their peckish impulses. "This suggests that the challenges of growing up in a large group drive the development of cognition," says Thornton. The cleverer female magpies also made better mothers, hatching more eggs and successfully raising more young. Moreover, in 2019 the team showed that large-group living in these birds sparks more innovative behaviors and facilitates their spread through social sharing of information.

The demands of an intricate social life may have spurred the evolution of intelligence in greater anis, too. For one thing, recognizing and getting to know their fellow group members is not an easy task. "These birds don't grow up together," says Riehl. "They're coming together as adults and learning to recognize one another—'Oh, you're in my group, but you are not'—which probably requires learned recognition of calls." The complex communication in the anis' chorusing displays also involves some sophisticated cognition, as does their collective decision-making.

How animals make decisions as a group is still a black box for science right now, says Riehl. "Different animals have different mechanisms for decision-making. Sometimes it's a dominant individual summarily making a choice and saying, 'Okay, you guys, follow me,' and sometimes it involves a group reaching consensus. Collective decision-making involves an ability to sense a threshold level of agreement, a quorum, before you can proceed. We don't know what the mechanism is in anis," says Riehl, "but surely it involves cognitive judgment: 'What's going on here?' 'Does everyone think this is a good spot or a bad spot for nesting?' That's a pretty hard thing to figure out." Riehl is just beginning to fashion cognitive tests for the greater ani to figure out how it might manage such sophisticated mental tasks.

Riehl sees roots of the greater ani's strange breeding system in the parasitic lifestyle of other cuckoos on its family tree. "But here's the cool thing," she says. "The anis act not only like brood parasites but also like their hosts. They're like conspecific brood parasites—they dump their eggs in the nests of other members of their species—but instead of abandoning their young, passing off parental care, the parasites stay at the nest and attend to their eggs and their nestlings." And like hosts, they eject eggs. Here's the answer to mystery number three. "If a female comes to the nest and finds an egg there before she has started laying—she sees that as a parasitic egg, so she gets rid of it," says Riehl. Greater anis don't like to have eggs in their nest before they've laid. It's like if you're expecting a baby, and you come home and find a baby in the crib—well, it's probably not yours. It's just as David Davis said, 'Anis are like parasites that have become host-specific on themselves.'"

The anis' extraordinary group display behavior—unique in the bird world—arose only after they had begun breeding communally.

In other words, the greater ani's way of breeding is evolving. "This is a very plastic behavior," says Riehl. "Right now, it's a mix of vestiges of parasitic behavior and adaptations to group living, like communal displays and collective decision-making. The birds haven't gotten to the point where they all just coordinate reproduction perfectly."

"That's the fun part for me," says Riehl, "thinking about how this weird mix of interesting behaviors evolved, what the original selective pressures were that favored them and how they changed over time so that something that was initially perhaps a parasitic behavior became something that was mutualistic and really social."

It's a whole different road to cooperative breeding than the "It's all about kinship" route championed for decades.

When Riehl surveyed cooperative breeding systems in hundreds of bird species, she found all varieties: birds such as fairy-wrens and Florida scrub jays that breed only in simple family groups (adults mate monogamously and helpers raise full siblings); birds such as greater anis and dunnocks that breed only with unrelated birds; birds that display everything in between, such as pied kingfishers found across Africa and Asia, which have all stripes of helpers at the nest, both kin and nonkin, and also superb starlings, birds that live in big, boisterous social groups of up to 40 individuals, mostly kin based but with multiple unrelated breeding females and a mix of related and unrelated breeding males. Of the 213 species Riehl looked at, 94 had complex alliances of kin and nonkin—most of them in Australia, Madagascar, and the neotropics, in other words, not in the Northern Hemisphere. Cooperative breeding groups made up of unrelated individuals are not nearly as rare as we once thought.

Another northern-bias-based assumption down the tubes.

And the reasons a stranger bird would join up with a motley group of nonkin? As various as the species that engage in it, says Riehl.

Sometimes birds have no choice. In gray jays, only the oldest fledging in a clutch gets to stay in his natal territory; he evicts all of his younger siblings, forcing them to settle with unrelated pairs.

Sometimes the cost of living alone is just too high. Amanda Ridley and her colleagues found that lone "floating" southern pied babblers struggle severely when they're out on their own. With no other babblers to serve as sentinels while they feed, they have to be more vigilant themselves. They forage less and lose weight. In the long term, floating is unsustainable, so they immigrate into groups as nonbreeding helpers.

Sometimes a lone bird—usually a male—will join a foreign pair or group, seeking to mate with the resident female. He'll often bring food, either for her or for her brood, in order to gain access. Then he'll hang around until the next breeding season, when his good behavior may allow him to inherit the breeding position. This is especially true for pied kingfishers, but also occurs with merlins, hoopoes, riflemen, and Puerto Rican todies. A lone male bell miner often becomes an especially hardworking helper at a nest, delivering more insects, fruit, and berries than related helpers, boosting the likelihood the female breeder will turn to him when her male partner dies. Whether or not she'll accept a helper as her mate depends on the amount of food he brings to her last clutch. Even if his chances are slim, this approach beats remaining with kin, where there's no chance at all of breeding.

As for why groups accept strangers: Anis and other cooperative breeders have taught us that collectives can raise young more effectively than lone pairs can and offer more vigilance and better protection from predators and brood parasites. (In 2013, scientists found that the global distribution of brood parasites and cooperative breeders overlapped closely, offering evidence that cooperative breeding may evolve as a defense against brood parasites.) Also, bigger groups displace smaller groups in competing for resources and nesting sites. And as we've seen, more birds make for a better ability to solve problems. In the end, the benefits of cooperative parental care outweigh the costs of sharing reproduction.

Sometimes the advantage—or the necessity—of a large group is so great that birds actually kidnap the young from other groups to boost their numbers. Cue the white-winged chough.

Robert Heinsohn says that choughs were his first love. One day, while he was searching for a topic for his PhD, he stumbled on the birds at a creek near the campus of Australian National University. "They were just acting so oddly," he says. "They were doing this spectacular and bizarre display, waving their wings and tails, flashing the white patches, dancing this little dance, and boggling their eyes—sort of engorging them with blood until they looked like they were bulging out of their skulls. One bird was displaying like this to a baby chough, a fledgling, and I thought, 'What is that?' And I ended up sitting with them for a couple of hours, just watching them."

Heinsohn was witnessing the first stage in the shocking act of chicknapping. It would take years of observation and research before he would see the whole picture.

Choughs are at the far end of the cooperative breeding spectrum, as obligate as you can get. They can't raise young without multiple helpers. This has to do with the way they forage, digging below the soil surface to get beetle larvae and earthworms. (It's a tricky method of feeding and takes skill developed over a long period of time, which is partly why choughs stay with their families for years.) Raising a single chick requires four birds working full time to scrape up enough food to feed it. For choughs, group living allows the birds to eke out an existence in a challenging environment. Even in good, wet years, the choughs are always trying to recruit other birds to their cause so they can get to that magical number of four, says Heinsohn. "They'll go to other groups of choughs and try to entice them over." But for the most part, it's all very peaceful. The birds spend their days methodically sifting leaf litter to search for food, with little squabbling or rivalry. Young birds stay at home to help their parents.

In years when drought strikes and the ground becomes hard and dry and food is difficult to find, it's another story. "Then all hell breaks loose," says Heinsohn. "Some of the older breeders die of starvation, and that causes complete social mayhem. The groups fragment, and some of the helpers take younger birds with them, forming little floating factions too small to breed on their own, temporary bands that get together for a while and then break up again. And the fragments are all flying around forming and reforming in a quest to find enough helpers."

That's when things turn really nasty. "The choughs start warring, destroying each other's nests," says Heinsohn. Larger groups mount raids on smaller groups, bullying and harassing them and stealing their newly fledged young.

The kidnapping begins with a baby waddling along on the ground, with its group all around it. When chough chicks fledge, they come out of the nest helpless and unable to fly. "They sort of half fly, half flop to the ground," says Heinsohn, so they're vulnerable to predators—and to abductors. "The groups are very cohesive, and they'll all be gathered around the baby. But if it goes off just a little bit on its own, choughs from other groups are watching, and one of the older helpers from another group will fly down to it and start to do that wing-wave, tail-wave display. The baby finds that display irresistible and gets caught up and follows the helper," says Heinsohn. From then on, it will work to feed the young in the new group. "I liken it to slave-making in ants that literally steal other ants and make them work. The young bird that's kidnapped from its family group ends up working for a group of birds that aren't related to it, and it gets nothing from the deal."

The chough story suggests one reason why cooperative breeding came to be. Maybe harsh, unpredictable environments favor the evolution of helping behavior—dubbed the "hard life" theory.

A slew of scientific papers has lately proposed this or that singular reason for cooperative breeding. "It's what evolutionary biologists live for," says Riehl, "to try to find the smoking gun, the common driver for things." Maybe cooperative breeding arose as protection from brood parasites or predators or unpredictable weather, acting as a buffer in bad years. Or maybe it originally evolved in good environments, where populations of paired species got so large, they saturated their habitats, outnumbering limited nest sites or food resources. Younger birds couldn't hold a territory of their own, so they opted to be helpers instead. Or maybe when birds live in good, stable habitats with plenty of rainfall and long growing seasons, they're more likely to transition from living in simple pairs to family living. With ample resources, it's easier for adults to keep their young around and for young to hang out and learn life skills from their parents. And then once they're living in groups, maybe exposure to harsh or changeable environments promotes the evolution of helping at the nest. Or maybe it's the other way around: Once they're established as groups, they're then able to invade habitats too harsh for independent breeders.

Riehl suspects that these different forces are at play in different kinds of cooperatively breeding birds. "If you think of cooperative breeding as one kind of behavior, then you look for one driver for that one kind of behavior," she says. "But if you think of it as an umbrella term that includes lots of different social systems that arose for different reasons, with different selective pressures, then there may be many drivers." Benefits of shared nesting range from spreading your own genes by helping your kin, to better reproductive success because of more food and more protection against predators and parasites. The costs are competition: Whose genes will benefit most from the efforts of the group? "It's not a complicated calculus," says Riehl. "But the particular costs and benefits could be different for different species."

L ook again at that bird family tree. You might imagine there is one branch that has settled on a shared way of being different from the other branches. But in fact, among bird groups, there are little twigs side by side, species closely related, that have followed paths to utterly different ways of being. The Florida scrub jay and the brown-headed nuthatch breed cooperatively; their close relatives, the California scrub jay and the red-breasted nuthatch, respectively, do not. The red-bellied woodpecker breeds in conventional pairs; its cousin, the acorn woodpecker, has one of the most complex communal nesting systems of any vertebrate. One species has evolved one life strategy; that close relative, that phylogenetic neighbor on a twig nearby, does something entirely different. Even a single species may diverge in its way of breeding. Among Australian magpies, the western population breeds cooperatively; in the east, the birds breed in pairs. And then there are little twigs on very different branches that have come up with similar ways of being by different paths and for different reasons. Birds as divergent as fork-tailed drongos and superb lyrebirds have both settled on mimicry as a useful life tool. Parrots and corvids, far, far apart on that family tree, have both evolved highly intelligent brains made for problem solving and for play.

It's what I love about birds. They're as inconsistent and unpredictable, as *varied*, as any group of animals on earth.

A LAST WORD

It's late August in the foothills of the Appalachians. The nesting season is nearly done, but there are still eastern bluebirds in pairs, flitting in and out of nesting boxes. Their way of raising young is so familiar here but so different from their western bluebird cousins, from acorn woodpeckers, scrub jays, and all the other species that nest as singletons or jointly as cooperative breeders.

In my backyard the other day, I saw a female cardinal singing. It was easy to spot her as the songster because of her species' dimorphic plumage. But I wonder about the times I may have seen a female chipping sparrow, cedar waxwing, or starling in song, indistinguishable from her male counterpart, and assumed it was a "he" singing.

Writing this book has changed the way I see birds, given me a new pair of binoculars, so to speak. Just for a day I'd like to experience the world the way they do, to see leaves with ultraviolet light baked into their greens, to hear and understand the minute musical differences and quick shifts in the acoustic structure of their complex calls and songs. Just for a day, I'd like to smell what a seabird smells, wake up one morning on the sea with a storm petrel's elevated olfactory sense and navigate the waters by swirling odor plumes and clouds. As the great essayist Lewis Thomas wrote, smell is a mode of knowing, remarkably like the act of thinking itself. The way birds use their senses offers clues to another kind of knowing.

Here's a reminder of the remarkable, enigmatic abilities of birds: I've just learned that nesting veeries are better than meteorologists at predicting the severity of an upcoming hurricane season. Ornithologist Christopher Heckscher discovered that veeries nesting in Delaware cut short their breeding season in years with the most numerous and intense hurricanes. Months in advance, they anticipate the storms and adjust their migratory schedules for crossing the Gulf of Mexico and the Caribbean Sea on their migration to South America to avoid the worst of the hurricanes. They also lay more eggs earlier in the season. How these birds know in May what will happen in August is a deep mystery, probably having to do with cues they pick up during their wintering season in South America. In their ability to predict the tropical storm season to come, the timing of nesting veeries is at least as good as—maybe a bit better than—the predictions by weather forecasters at the National Oceanic and Atmospheric Administration. Female veeries, that is. They're the ones that decide on the nesting and egg-laying schedule.

The fact that our old mistaken biases and assumptions are now giving way to more nuanced understanding of birds and their behavior is cause for optimism. Through new science, we've shed the belief that female birds are just along for the ride; that a bird is largely an eye driven by a wing, as if other senses were immaterial; that birds are by nature red in tooth and claw—their interactions with other individuals, other species, most often a matter of competition, of "scrunch or be scrunched." We now know that some birds are exceedingly cooperative. They collaborate in hunting everything from insects to rabbits, in putting on elaborate mating displays as "wingmen" with no clear advantage for themselves, and in raising the young of others. They use their voices to resolve conflict, negotiate boundaries, settle disputes, and spread the word about sources of food and danger. They take turns, in singing and in work and play.

Birds have taught us that classifying behavior into binary opposites—much as we like to do so—is often a futile exercise. Birds live and act on a spectrum, just as we do, and they prove the power of exceptions, both in defining rules and in breaking them.

Birds have also shown us that we're not unique in the ways we once thought. The teasing and clowning play in kea tells us that neither the capacity to be aware of other minds nor the wish to play with them belongs to humans alone. Nor are we alone in using aspects of language or tools, or in building complex structures, or in understanding, manipulating, and deceiving other animals. We may, however, be alone in devising reasons why we're special.

We understand now that birds are not just biologically distinct but culturally distinct—and that this is true even within a species. Birds learn styles of singing, bower building, playing that vary from one place, one population, to another. Palm cockatoos, kea, kelp gulls, great tits, superb fairy-wrens, and superb lyrebirds have taught us that birds use social learning to master different methods of foraging, to grasp the identity of their enemies, to master the dialect of their region or their own distinctive drumbeat.

Clearly there is no one way to be a bird, just as there is no one way to be a human. We have our different cultural ideals and shared practices across cultures, dynamic and changing; they have their individual identities and distinct behavioral and cultural practices, also dynamic—which they may share through social learning. But birds are all connected through the common thread of "birdness" just as we are all connected through our humanity.

To witness bird behavior in its full range is to glean some perspective on our own behavior. What happens with birds in periods of scarcity or ecological stress may not be so different from what happens with humans in times of environmental pressure, as Robert Heinsohn points out. For white-winged choughs, difficult ecological conditions such as drought tear the fabric of their society, setting the scene for societal collapse, power struggles, and complex Machiavellian social tactics—like kidnapping. Violence erupts. Some individuals triumph, but "in circumstances where it's almost impossible to live and breed without harassment," says Heinsohn. A social structure that functions well and supports the birds in good times is transformed into a harsh, violent, splintered system with big rewards only for the few, bullying of the smaller players, and hardship for most of the population.

With climate change, shrinking habitats, extinction of species, the analogy may become even more pertinent.

When I think about what we are doing to our fellow travelers on this planet, to the planet itself, I feel a wave of despair.

Just before his death in 2009, the world's leading expert on species associated with army ants, Carl Rettenmeyer, completed the first comprehensive list of animals known to keep company with *Eciton burchellii*, the "mini-lion" of neotropical forests. Rettenmeyer recorded 557 species in consort with the ants, ranging from mites and insects to a huge variety of birds. It's the largest group of animals ever described centering on a single species. At least 300 of these species depend on the ants, including the charismatic ocellated antbird. The disappearance of *Eciton burchellii* from any habitat over their vast range would trigger the extinction of hundreds of birds and other animals.

It can happen fast. Over the past decades, ornithologists have found that birds that depend on insects for sustenance are rapidly declining. Gone from European farmlands and countryside are little owls, beeeaters, Eurasian hobbies, and eight species of partridges. Populations of nightingales and turtledoves are greatly diminished. The culprit is not habitat destruction but starvation from the disappearance of their primary diet—beetles, dragonflies, and other insects. A 2019 study found that over the past twenty-five years, populations of birds that depend on insects fell by 13 percent across Europe and by almost 30 percent in Denmark.

That same year scientists delivered the shocking news that one in four birds in the US and Canada have disappeared since 1970—nearly three billion birds. The vanished species span the spectrum from meadowlarks, warblers, and swallows, to common backyard birds such as robins and sparrows. They're gone from all habitats, seashore, forest, grasslands, desert, tundra, probably due primarily to habitat loss from development and agriculture, as well as pesticide use. One recent study found that insecticides known as neonicotinoids prevent migratory birds from gaining the body mass and fat stores they need to start their journeys in a timely fashion. One can't help but think of Rachel Carson's prescient words in *Silent Spring*: "On the morning that had once throbbed with the dawn chorus of robins, catbirds, doves, jays, wrens, and scores of other bird voices, there was now no sound."

Also that year, bushfires swept Australia, killing more than a billion birds, mammals, and reptiles and destroying vast swaths of natural habitat.

Vulnerable to the ravages of climate change are more species than I can name, from migratory birds such as red knots susceptible to shifts in climate patterns that can upset the synchronization of their journeys, so delicately timed to coincide with the bloom of food sources, to seabirds with specialized diets such as petrels, which may face difficulties finding food to feed chicks because the abundance and distribution of their prey shifts radically in response to sudden environmental changes. A new study by the Woods Hole Oceanographic Institution warns that warming temperatures and disappearing sea ice may drive the extinction of emperor penguins in Antarctica by the end of this century.

Still, there are glimmers of hope. Say, in the resurgence of bald eagles and waterfowl and the puzzling but marvelous boost in vireo populations over the same period their warbler cousins have so dramatically declined, with eighty-nine million more vireos now than in 1970. Or in the surprising ability of a rainforest bird like the brush turkey to thrive in urban environments, or the little spotted antbird to fill a niche made vacant when the bigger, more dominant ocellated antbird goes extinct in its region. Or the way zebra finch parents inform their developing chicks of hot temperatures while they're still in the egg, and the chicks, in turn, adjust their growth and hatch smaller so they lose heat more easily to cool down. Or how the breeding behavior in many bird species seems flexible and fluid; perhaps certain species can beat the odds by shifting their breeding strategies. Or how social learning in tits and other birds allows information to spread faster, enabling them to adapt to new and challenging situations more quickly than evolution would allow. This sort of flexibility could give some species the ability to cope in a changing, more unpredictable world. Individual variation in birds may bring some resilience. It's the stuff of evolutionary change and also of innovation and keen problem solving.

On the drive back to Lund from his farm in Sweden, Mathias Osvath offered some perspective from his work with birds, only partly tongue-in-cheek. Corvids, he believes, are at the brink of a cognitive breakthrough. They have been around for millions of years. We humans have existed as a species for at most a few hundred thousand years—in geological terms, a flash in the pan, statistically insignificant. But in the short time of our existence, crows, ravens, and other corvids have learned to use us as a source of food and shelter. If our species disappears and corvids lose this resource, there may be selective pressure among them to boost cognition, says Osvath. Their brains could double or triple in size, and with their superefficient signaling and tight packing of neurons, they might become the next big thinkers, dominant among animals. It could happen very quickly, he says. "We dig up dinosaurs to try to figure out what happened to them. Perhaps someday dinosaurs in the form of corvids will dig us up to figure out what happened to us." He hopes they won't repeat our mistakes.

The word *auspicious*, meaning "favorable or promising success," comes from the Latin, *auspex*, or "observer of birds." In ancient Rome, bird-seers were priests, or augurs, who founded their divinations on the flight patterns of birds. The sixteenth-century English noun *auspice* originally referred to the practice of observing birds to find omens. There's a deep hunch here. We would do well to watch birds more, tune in to their usual and unusual behaviors, learn while we can from their marvelous—and still often mysterious—ways of being.

Acknowledgments

A book like this depends on the goodwill, generosity, and dedication of the naturalists and scientists who are doing brilliant work studying and observing birds in field and lab. They are the heroes for peeling back the layers of mystery and generating our new understanding of birds. To a person, all the researchers I've worked with have been unstinting in their willingness to give time and expertise and to share their enthusiasm for their work. I would like to thank especially those individuals who let me see their work in the field and took pains to give me a thorough introduction to their species.

My deep gratitude goes first to Andrew Skeoch, who responded to my simple query about a budgerigar photo with a very warm hello and an invitation to speak at a meeting of the Australian Wildlife Sound Recording Group in Baradine, Australia. En route to that gathering, he and his partner, potter Sarah Koschak, offered me kind hospitality and a front-seat view of the fairy-wrens in their yard, followed by a two-week tour of the landscapes, birds, and natural sounds of New South Wales. On that trip, Andrew, wizard of the acoustic landscape, opened my ears to a new way of listening to bird songs and calls and forever changed the way I hear them. He also read the entire manuscript and offered insightful suggestions.

Heartfelt thanks to Ana Dalziell, who took time away from work and family to show me her lyrebird research site in the heart of the Blue Mountains. Afterwards, she and Justin Welbergen sat with me answering my endless questions about the birds they study and showing me remarkable videos of lyrebird displays.

Deep thanks go to Tim Low, too, first for writing his masterful book about Australian birds, *Where Song Began*, and second, for spending a long day introducing me to the wild birds in and around his neighborhood. Not only did he show me the turkey mound in his backyard (which, he tells me, is now twice the size), but also gave me my first taste of Australian honeyeaters, magpies, cockatoos, bell miners, and dozens of other birds and offered his fascinating and learned perspective on them.

Will Feeney of Griffith University spent almost three full days with me at his field site, introducing me to the strange relationships between cuckoos and their hosts and his methods of studying them. My gratitude to him, as well as to the research team working with him to unravel the stories of these birds—James Kennerley, Maggie Grundler, Nicole Richardson, Derrick Thrasher, Joe Welklin, Julian Kapoor, Matthew Marsh, Rebecca Bracken, Zach Davis, Wendy Deptula, Stephanie LeQuier, Riley Neil, and Noah Hunt. Special thanks to James, Derrick, Joe, and Julian, who spent time with me describing their own interesting bird studies, past, present, and future.

Other Australian researchers who gave their time to show me their study sites or describe their captivating research include Michelle Hall of the University of Melbourne, Brani Igic, Darryl Jones of Griffith University, John Martin of the Royal Botanical Gardens in Sydney (who showed me my first powerful owl), and Robert Heinsohn, Naomi Langmore, and Robert Magrath of Australian National University.

Mary Caswell Stoddard spent many hours with me at her lab at Princeton University describing her enthralling work on avian color vision, eggs, and many other topics, and then kindly invited me to join her at the Rocky Mountain Biological Laboratory in Gothic, Colorado. There, she, Benedict Hogan, and Harold Eyster illuminated the wonders of the broad-tailed hummingbird dive display and the ingenious ways they are studying it. I'm deeply grateful for the time these talented young scientists spent with me.

Mathias and Helena Osvath welcomed me to their home and aviary near Lund, Sweden, and introduced me to their amazing ravens. Many thanks to the two of them and to Mathias for so generously sharing his vast knowledge of these birds and of avian play in general. Deep gratitude also to Raoul Schwing and Amelia Wein for giving me a spirited introduction to kea and a window into the inspired work they do to understand this most mischievous, endearing, and playful bird.

A number of scientists gave their time to talk with me in depth about their research, among them: James Dale of Massey University, Emily DuVal of Florida State University, Van Graham of Wildlife West, Jessica McLachlan of

Cambridge University, Angela Medina-García of the University of Colorado, Boulder, Sean McCann of Arcadia University, Gabrielle Nevitt of the University of California, Davis, Sean O'Donnell of Drexel University, Juan Reboreda of Universidad de Buenos Aires, Christina Riehl of Princeton University, Alex Taylor of the University of Auckland, biologist Paul Tebbel, Christopher Templeton of Pacific University, Michael Webster of the Cornell Lab of Ornithology, and Tim Wright of New Mexico State University.

I owe a large debt of gratitude to Sue Healy of the University of St. Andrews for not only taking the time to discuss with me her research and thoughts on hummingbirds, nest building, and wild cognition but for reading the entire manuscript with care and thoughtfulness and offering helpful suggestions.

Many other busy but generous individuals corresponded with me at length about their research, supplied references, and read and reread drafts of passages in the book exploring their work, including all of the abovementioned scientists and naturalists, as well as Nick Ackroyd and Kirsty Wills of the Willowbank Wildlife Reserve; Phil Battley of Massey University; Yitzchak Ben Mocha of the Max Planck Institute for Ornithology; ethno-ornithologist Mark Bonta; Gerald Borgia of the University of Maryland; Patricia Brennan of Mt. Holyoke College; Signe Brinkløv of the University of Southern Denmark; Thomas Bugnyar of the University of Vienna; Johel Chaves-Campos of the Council on International Educational Exchange, Study Abroad Program in Tropical Ecology and Conservation in Costa Rica; John Endler of Deakin University; Sabrina Engesser of the University of Vienna; ornithologist Robert Gosford; George Happ of the University of Alaska and the University of Vermont; Christopher Heckscher of Delaware State University; Suzana Herculano-Houzel of Vanderbilt University; Jason Keagy of the University of Illinois at Urbana-Champaign; Laura Kelley of the University of Exeter; Matthew I. M. Louder of the University of Tokyo; Jessica Meir of NASA; Julien Meyer of the University of Grenoble; Kees Moeliker of the Natural History Museum in Rotterdam; Ximena Nelson of the University of Canterbury; Karan Odom of the Cornell Laboratory of Ornithology; Paul Ponganis of the University of California, Santa Cruz; Simon Potier of Lund University; Ed Scholes of the Cornell Laboratory of Ornithology; Jonathan Slaght of the Wildlife Conservation Society; Claire Spottiswoode of the University of Cape Town and the University of Cambridge; Diego Sustaita of California State University, San Marcos; Toshitaka Suzuki of Kyoto University; Christopher Templeton of Pacific University;

Alex Thornton of the University of Exeter; Julia York of the University of Texas, Austin; zoologist and photographer Christy Yuncker; and Marlene Zuk of the University of Minnesota.

The comments and contributions of all of these scientists enriched the book enormously and were invaluable in ensuring its accuracy. If mistakes persist, they are mine alone.

Special thanks to the members of the Australian Wildlife Sound Recording Group for sharing with me the marvelous endeavor of recording natural sounds and their own experiences with Australian birds and other wildlife at their annual meeting in Baradine in 2017. For opening my ears, I'd like to thank in particular Ros Bandt, Leah Barclay, Tony Baylis, Jessie Cappadonna, Lucy Farrow, Sue Gould, Vicki Hallett, Michael Mahoney, Bob Tomkins, Andrew Skeoch, and Fred Van Gessel. I will never forget sitting silently by that little bush dam on our final evening and watching, enchanted, as a dozen glossy black cockatoos descended to drink.

Thanks, too, to Ian Billick of the Rocky Mountain Biological Laboratory, for the introduction to RMBL, and to the organizers and participants of the many bird festivals I attended who offered bird walks and talks and introductions to fascinating species and people. I especially appreciated the efforts of Abel Julien, Francesc Kirchner, and Miquel Rafa of the Delta Birding Festival in Catalonia, Spain; Nancy Merrill of the Yampa Valley Crane Festival; Jonathan Weston of the Roger Tory Peterson Wild America Nature Festival; Bob Elner and Rob Butler of the Vancouver International Bird Festival; and Mallory Primm and Robbi Mixon of the Kachemak Bay Shorebird Festival. Deep appreciation also goes to Albin Grahn for a wonderful expedition to Angarnsjöängen in Sweden to see the Eurasian coots, gray herons, Eurasian cranes, hen harriers, and a young white-tailed eagle, and for giving me that magical moment with Percy, the great gray owl at Skansen.

I owe gratitude to the many friends and colleagues who offered support, enlightenment, sustenance, ideas, photos, and birding adventures, real and virtual—among them, Susan Allison, Susan Bacik, Ros Casey, Kathy Clark, Daniel Delano, Laura Delano, Mark Edmundson, Andrew Floyd, Ted Floyd, Greg Gelburd, Dorrit Green, Robin Hanes, Roger Hirschland, Susan Hitchcock, Brian Hofstetter (who sent me his own wonderful list of unusual bird behaviors), Rich Hofstetter, Johanna Holldack, Meg Jay, Karen B. London,

Donna Lucey, Michele Martin, Leslie Middleton, Nancy Murphy-Spicer, Miriam Nelson, Debra Nystrom, Dan O'Neill, Diane Ober, Michael Rodemeyer, Sandy Schmidt, David Spicer, Ellen and Paul Wagner, Jo White, Henry Wiencek, and Andrew Wyndham. Special thanks to Sandy Cushman, Liz Denton, Sharon Hirschland, and Pete Myers for a steady stream of engaging and illuminating conversations and walks—and again to Pete for so generously sharing his brilliant photos.

I want to give thanks for my dear friend Katherine Magraw, who died of cancer in April 2019. Katherine was a brave and loving force for good in this world. I feel deeply fortunate to have had her as my friend and to know and love her amazing daughters, Emma and Grace, who brought to our household warmth, humor, and light.

The three years during which I wrote this book were challenging, and I am lucky to have the close family I have. Thanks to Hoey and Ronnie for showing up in my life, in the Colorado mountains, and in my yard when I most needed their strong, able assistance and kind company; to Nan, for leaning in with offers of medical help, sympathetic words, and a steady flow of invitations; to John, for assisting my extended family with bottomless patience, creativity, and spirit; to my adored sisters, Kim, Nancy, and Sarah, for standing by me; and to dear Gail and Bill for their generous love and support.

To Nance and Steve, I feel endless gratitude every day for their essential, loving, generous, and sustaining presence in my life.

I owe more thanks than I can say to my agent, Melanie Jackson—for her shining intelligence and warm support in difficult times, and for her always, always pitch-perfect editorial perspective and advice. Melanie led me to Ann Godoff, the editor every writer hopes for. Thank you, Ann, for your extraordinary wisdom, your clarity, your grace. Thanks to Casey Denis for her expert and creative help in moving the manuscript through the publication process, to Darren Haggar for the striking jacket design, to Eunike Nugroho for the exquisite artwork illuminating the cover, and to John Burgoyne for the lovely illustrations throughout the book.

Finally, my profound gratitude to my daughters, Zoë and Nelle. Without their words of affirmation, playlists, cartoons, household help, camaraderie, humor, perspective, advice, honesty, love, and support, this book would not be. To Nelle I owe special thanks for taking on the formal role of editorial assistant and performing the job brilliantly and patiently and for carefully transcribing dozens of recorded interviews, organizing speaking events, planning and arranging travel and itineraries, and, above all, proffering her keen editorial eye and always sage advice. Thank you for your abundant help, Nelle, given even in the midst of overwhelming challenges to your health and well-being. You are owl, ani, hawk, raven, hummingbird, and veery rolled into one.

Further Reading

INTRODUCTION: WHEN YOU'VE SEEN ONE BIRD

G. F. Barrowclough, et al., "How many kinds of birds are there and why does it matter?" *PLOS ONE* 11, no. 11 (2016): doi/10.1371/journal.pone.0166307.

N. J. Boogert et al., "Measuring and understanding individual differences in cognition," *Philosophical Transactions of the Royal Society B* 373 (2018): 20170280.

J. Dale, "Plumage Color in Males and Females" (keynote lecture, 54th Annual Conference of the Animal Behavior Society, Toronto, 2017).

R. E. Gill et al., "Extreme endurance flights by landbirds crossing the Pacific Ocean: Ecological corridor rather than barrier?" *Proceedings of the Royal Society B: Biological Sciences* 276, no. 1656 (October 2008): 447–57.

J. Gould, *The Birds of Australia*, 7 vols. (London: Richard and John E. Taylor, 1848).

D. Griffin, *Animal Thinking* (Cambridge, MA: Harvard University Press, 1984).

S. D. Healy, "Animal cognition," *Integrative Zoology* 14, no. 2 (2019): 128–31.

S. D. Healy et al., "Explanations for variation in cognitive ability: Behavioural ecology meets comparative cognition," *Behavioural Processes* 80, no. 3 (2009): 288–94.

R. Heinsohn, "Ecology and evolution of the enigmatic eclectus parrot (*Eclectus roratus*)," *Journal of Avian Medicine and Surgery* 22, no. 2 (2008): 146–50.

R. Heinsohn, "Eclectus' True Colors Revealed," *Bird Talk*, February 2009, 38.

R. Heinsohn et al., "Extreme reversed sexual dichromatism in a bird without sex role reversal," *Science* 22, no. 309 (2005): 617–19.

S. Herculano-Houzel, "Numbers of neurons as biological correlates of cognitive capacity," *Current Opinion in Behavioral Sciences* 16 (2017): 1–7.

L. Lefebvre, "Taxonomic counts of cognition in the wild," *Biology Letters* 7, no. 4 (August 2011): 631–33.

T. Low, *Where Song Began: Australia's Birds and How They Changed the World* (New Haven, CT: Yale University Press, 2016).

J. U. Meir, "Physiology at the Extreme: From Ocean Depths to Mountain Peaks Among the Stars" (plenary lecture, North American Ornithological Conference, Washington, DC, August 17, 2016).

J. U. Meir and W. K. Milsom, "High thermal sensitivity of blood enhances oxygen delivery in the high-flying bar-headed goose," *Journal of Experimental Biology* 216 (2013): 2172–75.

J. U. Meir et al., "Heart rate and metabolic rate of bar-headed geese flying in hypoxia," *Federation of American Societies for Experimental Biology Journal* 27 (2013).

J. U. Meir et al., "Reduced metabolism supports hypoxic flight in the high-flying bar-headed goose (*Anser indicus*)," *eLife* 8 (2019): e44986.

K. J. Odom et al., "Female song is widespread and ancestral in songbirds," *Nature Communications* 5 (2014): 3379.

D. J. Pritchard et al., "Why study cognition in the wild (and how to test it)?," *Journal of the Experimental Analysis of Behavior* 105, no. 1 (2016): 41–55.

D. J. Pritchard et al., "Wild rufous hummingbirds use local landmarks to return to rewarded locations," *Behavioural Processes* 122 (2016): 59–66.

K. Riebel et al., "New insights from female bird song: Towards an integrated approach to studying male and female communication roles," *Biology Letters* 15, no. 4 (2019): doi/full/10.1098/rsbl.2019.0059.

L. Robin, R. Heinsohn, and L. Joseph, eds., *Boom & Bust: Bird Stories for a Dry Country* (Canberra, Australia: CSIRO Publishing, 2009).

G. R. Scott et al., "How bar-headed geese fly over the Himalayas," *Physiology 30*(2) (2015): 107–15.

S. Shaffer discovery reported in the *East Bay Times*: P. Rogers, "Hitchhiking Gull Takes 150-Mile Truck Ride Along California Freeways," July 13, 2018, https://www.eastbaytimes.com/2018/07/13 /hitchhiking-seagull-takes-150-mile-truck-ride-along-california-freeways/.

L. Swan, *Tales of the Himalaya: Adventures of a Naturalist* (La Crescenta, CA: Mountain N' Air Books, 2000), 90.

J. F. Welklin, "Neighborhood bullies: The importance of social context on plumage in redbacked fairy-wrens" (Sigma Xi Mini-Symposium, Cornell University, Ithaca, NY, February 2015).

J. F. Welklin et al., "Social environment, costs, and the evolution of sexual signals" (EvoDay Symposium, Cornell University, Ithaca, NY, May 8, 2015).

TALK

CHAPTER ONE: DAWN CHORUS

K. S. Berg et al., "Phylogenetic and ecological determinants of the neotropical dawn chorus," *Proceedings of the Royal Society B: Biological Sciences* 273, no. 1589 (2006): 999–1005.

D. Colombelli-Négrel et al., "Embryonic learning of vocal passwords in superb fairy-wrens reveals intruder cuckoo nestlings," *Current Biology* 22 (2012): 2155–60.

D. Colombelli-Négrel et al., "Prenatal learning in an Australian songbird: Habituation and individual discrimination in superb fairy-wren embryos," *Proceedings of the Royal Society B: Biological Sciences* 281, no. 1797 (2014): 20141154.

J. Dale, "Ornamental plumage does not signal male quality in red-billed queleas," *Proceedings of the Royal Society B: Biological Sciences* 267, no. 1458 (2000): 2143–49.

A. H. Dalziell and A. Cockburn, "Dawn song in superb fairy-wrens: A bird that seeks extrapair copulations during the dawn chorus," *Animal Behaviour* 75, no. 2 (2008): 489–500.

K. Delhey et al., "Cosmetic coloration in birds: Occurrence, function, and evolution," supplement, *American Naturalist* 169, no. S1 (2007): S145–S158.

C. Dreifus, "Luis Baptista, 58, an Author and an Expert on Bird Song," *The New York Times*, June 27, 2000).

G. Happ, *Sandhill Crane Display Dictionary: What Cranes Say with Their Body Language* (Dunedin, FL: Waterford Press, 2017).

S. Hoffmann et al., "Duets recorded in the wild reveal that interindividually coordinated motor control enables cooperative behavior," *Nature Communications* 10, no. 2577 (2019): doi:10.1038/s41467-019 -10593-3.

S. Keen et al., "Song in a social and sexual context: Vocalizations signal identity and rank in both sexes of a cooperative breeder," *Frontiers in Ecology and Evolution* 4, no. 46 (2016): doi:10.3389/fevo .2016.00046.

E. Kemmerer et al., "High densities of bell miners *Manorina melanophrys* associated with reduced diversity of other birds in wet eucalypt forest: Potential for adaptive management," *Forest Ecology and Management* 255, no. 7 (2008): 2094–2102.

D. M. Logue and D. B. Krupp, "Duetting as a collective behavior," *Frontiers in Ecology and Evolution* (2016): doi.org/10.3389/fevo.2016.00007.

B. Mampe et al., "Newborns' cry melody is shaped by their native language," *Current Biology* 19, no. 23 (2009): 1994–97.

M. M. Mariette et al., "Parent-embryo acoustic communication: A specialised heat vocalisation allowing embryonic eavesdropping," *Scientific Reports* 8, no. 10 (2018): 17721.

J. P. Myers, "One deleterious effect of mobbing in the southern lapwing *(Vanellus chilensis)*," *Auk* 95, no. 2 (1978): 419–20.

S. A. Nesbitt, "Feather staining in Florida sandhill cranes," *Florida Field Naturalist* (fall 1975): 28–30.

C. Pérez-Granados et al., "Dawn chorus interpretation differs when using songs or calls: The Dupont's lark *Chersophilus duponti* case," *Peer Journal* 6 (2018): e5241.

K. D. Rivera-Cáceres, "The Ontogeny of Duets in a Neotropical Bird, the Canebrake Wren" (PhD diss., University of Miami, 2017), Open Access Dissertations, 1830.

K. D. Rivera-Cáceres and C. N. Templeton, "A duetting perspective on avian song learning," *Behavioural Processes* 163 (2019): 71–80.

K. D. Rivera-Cáceres et al., "Early development of vocal interaction rules in a duetting songbird," *Royal Society Open Science* 5, no. 2 (2018): 171791.

A. C. Rogers et al., "Function of pair duets in the eastern whipbird: Cooperative defense or sexual conflict?," *Behavioral Ecology* 18, no. 1 (2007): 182–88.

O. Tchernichovski et al., "How social learning adds up to a culture: From birdsong to human public opinion," *Journal of Experimental Biology* 220 (2017): 124–32.

C. N. Templeton et al., "An experimental study of duet integration in the happy wren, *Pheugopedius felix*," *Animal Behaviour* 86, no. 4 (2013): 821-27.

R. J. Thomas et al., "Eye size in birds and the timing of song at dawn," *Proceedings of the Royal Society B: Biological Sciences* 269 (2002): 831–37.

J. A. Tobias et al., "Territoriality, social bonds, and the evolution of communal signaling in birds," *Frontiers in Ecology and Evolution* 24 (2016): doi: 10.3389/fevo.2016.00074.

CHAPTER TWO: CAUSE FOR ALARM

B. E. Byers and D. E. Kroodsma, "Avian Vocal Behavior," in *The Cornell Lab of Ornithology Handbook of Bird Biology*, eds. I. J. Lovette and J. W. Fitzpatrick, 3rd ed. (Hoboken, NJ: Wiley, 2016): 355–405.

S. S. Cunningham and R. D. Magrath, "Functionally referential alarm calls in noisy miners communicate about predator behaviour," *Animal Behaviour* 129 (2017): 171–79.

E. Curio et al., "Cultural transmission of enemy recognition: One function of mobbing," *Science* 202, no. 4370 (1978): 899–901.

F. S. E. Dawson Pell, "Birds orient their heads appropriately in response to functionally referential alarm calls of heterospecifics," *Animal Behaviour* 140 (2018): 109–18.

S. Engesser et al., "Chestnut-crowned babbler calls are composed of meaningless shared building blocks," *PNAS* 116, no. 39 (2019): 19579–84: doi.org/10.1073/pnas.1819513116.

S. Engesser et al., "Experimental evidence for phonemic contrasts in a nonhuman vocal system," *PLOS Biology* 13, no. 6 (2015): e1002171.

S. Engesser et al., "Internal acoustic structuring in pied babbler recruitment cries specifies the form of recruitment," *Behavioral Ecology* 29, no. 5 (2018): 1021–30.

M. Hingee and R. D. Magrath, "Flights of fear: A mechanical wing whistle sounds the alarm in a flocking bird," *Proceedings of the Royal Society B: Biological Sciences* 276, no. 1676 (2009): 4173–79.

B. Igic et al., "Crying wolf to a predator: Deceptive vocal mimicry by a bird protecting young," *Proceedings of the Royal Society B: Biological Sciences* 282, no. 1809 (2015): doi.org/10.1098/rspb.2015.0798.

B. Jones, "Long-lasting cognitive and behavioral effects of single encounter with predator," (presentation, International Ornithological Congress, Vancouver, August 26, 2018).

S. L. Lima and L. M. Dill, "Behavioral decisions made under the risk of predation: A review and prospectus," *Canadian Journal of Zoology* 68, no. 4 (1990): 619–40.

R. D. Magrath and T. H. Bennett, "A micro-geography of fear: Learning to eavesdrop on alarm calls of neighbouring heterospecifics," *Proceedings of the Royal Society, B: Biological Sciences* 279, no. 1730 (2012): 902–09.

R. D. Magrath et al., "An avian eavesdropping network: Alarm signal reliability and heterospecific response," *Behavioral Ecology* 20, no. 4 (2009): 745–52.

R. D. Magrath et al., "Eavesdropping on heterospecific alarm calls: From mechanisms to consequences," *Biological Reviews* 90, no. 2 (2015): 560–86.

R. D. Magrath et al., "Recognition of other species' aerial alarm calls: Speaking the same language or learning another?," *Proceedings of the Royal Society B: Biological Sciences* 276, no. 1657 (2009): 769–74.

R. D. Magrath et al., "Wild birds learn to eavesdrop on heterospecific alarm calls," *Current Biology* 25, no. 15 (2015): 2047–50.

J. R. McLachlan, "Alarm Calls and Information Use in the New Holland Honeyeater" (PhD thesis, University of Cambridge, 2019).

J. R. McLachlan, "How an alarm signal encodes for when to flee and for how long to hide," symposium at the International Ornithological Congress, Vancouver, August 26, 2018.

J. Meyer and D. D. Reyes, "Geolingüística de los lenguajes silbados del mundo, con un enfoque en el español silbado," *Géolinguistique* 17 (2017): 99–124.

T. G. Murray et al., "Sounds of modified flight feathers reliably signal danger in a pigeon," *Current Biology* 27, no. 22 (2017): P3520-3525.E4.

D. A. Potvin et al., "Birds learn socially to recognize heterospecific alarm calls by acoustic association," *Current Biology* 28 (2018): 2632–37.

R. M. Seyfarth et al., "Vervet monkey alarm calls: Semantic communication in a free-ranging primate," *Animal Behaviour* 28, no. 4 (1980): 1070–94.

R. M. Seyfarth et al., "Monkey responses to three different alarm calls: Evidence of predator classification and semantic communication." *Science* 210, no. 4471 (1980): 801-803.

T. N. Suzuki, "Semantic communication in birds: Evidence from field research over the past two decades," *Ecological Research* 31, no. 3 (2016): 307–19.

T. N. Suzuki et al., "Wild birds use an ordering rule to decode novel call sequences," *Current Biology* 27, no. 15 (2017): 2331–36.

C. N. Templeton et al., "Allometry of alarm calls: Black-capped chickadees encode information about predator size," *Science* 308, no. 5730 (2005): 1934–37.

CHAPTER THREE: SUPERB PARROTING

R. W. Byrne and N. Corp, "Neocortex size predicts deception rate in primates," *Proceedings of the Royal Society B: Biological Sciences* 271, no. 1549 (2004): 1693–99.

T. Caro, "Antipredator deception in terrestrial vertebrates," *Current Zoology* 60, no. 1 (2014): 16–25.

A. H. Chisholm, *Nature's Linguists: A Study of the Riddle of Vocal Mimicry* (Burwood, Australia: Brown, Prior, Anderson, 1946).

A. H. Dalziell, "Avian vocal mimicry: A unified conceptual framework," *Biological Reviews* 90, no. 2 (2015): 643–58.

A. H. Dalziell and R. D. Magrath, "Fooling the experts: Accurate vocal mimicry in the song of the superb lyrebird, *Menura novaehollandiae*," *Animal Behaviour* 83, no. 6 (2012): 1401–10.

A. H. Dalziell and J. A. Welbergen, "Elaborate mimetic vocal displays by female superb lyrebirds," *Frontiers in Ecology and Evolution* (2016): doi.org/10.3389/fevo.2016.00034.

A. H. Dalziell et al., "Dance choreography is coordinated with song repertoire in a complex avian display," *Current Biology* 23, no. 12 (2013): 1132–35.

N. J. Emery and N. S. Clayton, "The mentality of crows: Convergent evolution of intelligence in corvids and apes," *Science* 306, no. 5703 (2004): 1903–07.

T. P. Flower et al., "Deception by flexible alarm mimicry in an African bird," *Science* 344, no. 6183 (2014): 513–16.

M. Goller and D. Shizuka, "Evolutionary origins of vocal mimicry in songbirds," *Evolution Letters* 2, no. 4 (2018): 417–26.

V. A. Gombos, "The cognition of deception: The role of executive processes in producing lies," *Genetic, Social, and General Psychology Monographs* 132, no. 3 (2006): 197–214.

B. Igic and R. D. Magrath, "Fidelity of vocal mimicry: Identification and accuracy of mimicry of heterospecific alarm calls by the brown thornbill," *Animal Behaviour* 85, no. 3 (2013): 593–603.

B. Igic and R. D. Magrath, "A songbird mimics different heterospecific alarm calls in response to different types of threat," *Behavioral Ecology* 25, no. 3 (2014): 538–48.

A. C. Katsis et al., "Prenatal exposure to incubation calls affects song learning in the zebra finch," *Scientific Reports* 8 (2018): 15232.

L. A. Kelley and S. D. Healy, "Vocal mimicry in male bowerbirds: Who learns from whom?," *Biology Letters* 6, no. 5 (2010): 626–29.

J. F. Prather et al., "Precise auditory-vocal mirroring in neurons for learned vocal communication," *Nature* 451, no. 7176 (2008): 305–10.

D. A. Putland et al., "Imitating the neighbours: vocal dialect matching in a mimic-model system," *Biology Letters* 2, no. 3 (2006): 367–70.

R. M. Sapolsky, *Behave: The Biology of Humans at Our Best and Worst* (New York: Penguin Press, 2017).

R. A. Suthers and S. A. Zollinger, "Producing song: The vocal apparatus," *Annals of the New York Academy of Sciences* 1016 (2004): 109–29.

R. Zann and E. Dunstan, "Mimetic song in superb lyrebirds: Species mimicked and mimetic accuracy in different populations and age classes," *Animal Behaviour* 76, no. 3 (2008): 1043–54.

WORK

CHAPTER FOUR: THE SCENT OF SUSTENANCE

J. J. Audubon, "Account of the habits of the turkey buzzard, Vultur aura, particularly with the view of exploding the opinion generally entertained of its extraordinary power of smelling," *Edinburgh New Philosophical Journal* 2 (1826): 172–84.

B. G. Bang, "Anatomical adaptations for olfaction in the snow petrel," *Nature* 205 (1965): 513–15.

B. G. Bang, "Anatomical evidence for olfactory function in some species of birds," *Nature* 188 (1960): 547–49.

B. G. Bang, "The olfactory apparatus of tubenosed birds (Procellariiformes)," *Acta Anatomica* 65, no. 1 (1966): 391–415.

D. Bakaloudis, "Hunting strategies and foraging performance of the short-toed eagle in the Dadia-Lefkimi-Soufli National Park, north-east Greece," *Journal of Zoology* 281, no. 3 (2010): 168–74.

F. Bonadonna et al., "Evidence that blue petrel, *Halobaena caerulea*, fledglings can detect and orient to dimethyl sulfide," *Journal of Experimental Biology* 209 (2006): 2165–69.

S. Brinkløv et al., "Echolocation in oilbirds and swifts," *Frontiers in Physiology* 4 (2013): 123.

S. Brinkløv et al., "Oilbirds produce echolocation signals beyond their best hearing range and adjust signal design to natural light conditions," *Royal Society Open Science* 4, no. 5 (2017): 170255.

G. C. Cunningham and G. A. Nevitt, "Evidence for olfactory learning in procellariiform seabird chicks," *Journal of Avian Biology* 42, no. 1 (2011): 85–88.

S. J. Cunningham and I. Castro, "The secret life of wild brown kiwi: Studying behaviour of a cryptic species by direct observation," *New Zealand Journal of Ecology* 35, no. 3 (2011): 209–19.

J. L. DeBose and G.A. Nevitt, "The use of odors at different spatial scales: Comparing birds with fish," *Journal of Chemical Ecology* 34, no. 7 (2008): 867–81.

G. De Groof et al., "Neural correlates of behavioural olfactory sensitivity changes seasonally in European starlings," *PLOS ONE* 5, no. 12 (2010): e14337.

P. Estók et al., "Great tits search for, capture, kill and eat hibernating bats," *Biology Letters* 6, no. 1 (2010): 59–62.

D. R. Griffin, "Acoustic orientation in the oil bird, *Steatornis*," *Proceedings of the National Academy of Sciences* 39, no. 8 (1953): 884–93.

D. R. Griffin, "How I Managed to Explore the 'Magical' Sense of Bats," *Scientist*, October 3, 1988.

N. P. Grigg et al., "Anatomical evidence for scent guided foraging in the turkey vulture," *Scientific Reports* 7 (2017): 17408.

H. Gwinner et al., "Green plants in starling nests: Effects on nestlings. *Animal Behaviour* 59 (2010): 301–9.

J. C. Hagelin and I. L. Jones, "Bird odors and other chemical substances: A defense mechanism or overlooked mode of intraspecific communication?," *Auk* 124, no. 3 (2007): 741–61.

K. A. Hindwood, "A feeding habit of the shrike-tit," *Emu* 46 (1946): 284-85.

R. A. Holland et al., "The secret life of oilbirds: New insights into the movement ecology of a unique avian frugivore," *PLOS ONE* 4, no. 12 (2009): e8264.

G. R. Martin et al., "The eyes of oilbirds (*Steatornis caripensis*): Pushing at the limits of sensitivity," *Naturwissenschaften* 91, no. 1 (2004): 26–29.

R. Montgomerie and P. J. Weatherhead, "How robins find worms," *Animal Behaviour* 54, no. 1 (1997): 143–51.

G. A. Nevitt, "Sensory ecology on the high seas: The odor world of the procellariiform seabirds," *Journal of Experimental Biology* 211 (2008): 1706–13.

G. A. Nevitt and J. C. Hagelin, "Symposium overview: Olfaction in birds: A dedication to the pioneering spirit of Bernice Wenzel and Betsy Bang, *Annals of the New York Academy of Sciences* 1170, no. 1 (2009): 424–27.

G. A. Nevitt et al., "Evidence for olfactory search in wandering albatross, *Diomedea exulans*," *Proceedings of the National Academy of Sciences* 105, no. 12 (2008): 4576–81.

R. S. Payne, "Acoustic location of prey by barn owls (*Tyto alba*)," *Journal of Experimental Biology* 54 (1971): 535–73.

S. Potier et al., "Sight or smell: Which senses do scavenging raptors use to find food?," *Animal Cognition* 22, no. 1 (2019): 49–59.

J. C. Slaght et al., "Global Distribution and Population Estimate of Blakiston's Fish Owl," in *Biodiversity Conservation Using Umbrella Species: Blackiston's Fish Owl and the Red-Crowned Crane*, ed. F. Nakamura (New York: Springer, 2018): 9–18.

J. C. Slaght et al., "Ecology and Conservation of Blakiston's Fish Owl in Russia," in *Biodiversity Conservation Using Umbrella Species: Blackiston's Fish Owl and the Red-Crowned Crane*, ed. F. Nakamura (New York: Springer, 2018): 47–70.

K. E. Stager, "The role of olfaction in food location by the turkey vulture (*Cathartes aura*)," *Los Angeles County Museum Contributions in Science* 81 (1964): 3–63.

M. S. Stoddard and R. Prum, "Evolution of avian plumage color in a tetrahedral color space: A phylogenetic analysis of New World buntings," *American Naturalist* 171, no. 6 (2008): 755–76.

D. Sustaita et al., "Come on baby, let's do the twist: The kinematics of killing in loggerhead shrikes," *Biology Letters* 14, no. 9 (2018).

C. Tedore and D.-E. Nilsson, "Avian UV vision enhances leaf surface contrasts in forest environments," *Nature Communications* 10 (2019): 239.

R. W. Van Buskirk and G. A. Nevitt, "The influence of developmental environment on the evolution of olfactory foraging behavior in procellariiform seabirds," *Journal of Evolutionary Biology* 21, no. 1 (2008): 67–76.

A. von Humboldt, *Personal Narrative of a Journey to the Equinoctial Regions of the New Continent: Abridged Edition*, trans. and ed. Jason Wilson (New York: Penguin Classics, 1996).

H. Weimerskirch et al., "Use of social information in seabirds: Compass rafts indicate the heading of food patches," *PLOS ONE* 5, no. 3 (2010): e9928.

S.-Y. Yang et al., "Stop and smell the pollen: The role of olfaction and vision of the oriental honey buzzard in identifying food," *PLOS ONE* 10, no. 7 (2015): e0130191.

CHAPTER FIVE: HOT TOOLS

L. Aplin, "Culture and cultural evolution in birds: A review of the evidence," *Animal Behaviour* 147 (2019): 179–87.

P. Barnard, "Foraging site selection by three raptors in relation to grassland burning in a montane habitat," *African Journal of Ecology* 25, no. 1 (1987): 35–45.

M. Bonta et al., "International fire-spreading by 'firehawk' raptors in northern Australia," *Journal of Ethnobiology* 37, no. 4 (2017): 700–18.

N. J. Emery and N. S. Clayton, "Effects of experience and social context on prospective caching strategies by scrub jays," *Nature* 414, no. 6862 (2001): 443–46.

D. C. Gayou, "Tool use by green jays," *Wilson Bulletin* 94 (1982): 595–96.

C. Green, "Use of tool by orange-winged sitella," *Emu* 71, no. 1 (1972): 185–86.

R. Gruber et al., "New Caledonian crows use mental tool representations to solve metatool problems," *Current Biology* 29, no. 4 (2019): 686–92.

J. N. Hobbs, "Use of tools by the white-winged chough," *Emu* 71, no. 2 (1971): 84–85.

T. J. Hovick et al., "Pyric-carnivory: Raptor use of prescribed fires," *Ecology and Evolution* 7, no. 21 (2017): 9144–50.

B. Kenward et al., "Development of tool use in New Caledonian crows: Inherited action patterns and social influences," *Animal Behaviour* 72, no. 6 (2006): 1329–43.

J. S. Marks and C. S. Hall, "Tool use by bristle-thighed curlews feeding on albatross eggs," *Condor* 94, no. 4 (1992): 1032–34.

F. F. Marón et al., "Increased wounding of southern right whale (*Eubalaena australis*) calves by kelp gulls (*Larus dominicanus*) at Península Valdés, Argentina," *PLOS ONE* 10, no. 10 (2015): 1–20.

A. Skutch, *The Minds of Birds* (College Station, TX: Texas A&M University Press, 1996).

A. M. P. von Bayern et al., "Compound tool construction by New Caledonian crows," *Scientific Reports* 8, no. 1 (2018): 15676.

CHAPTER SIX: TRACING THE ANT'S PATH

M. Araya-Salas et al., "Spatial memory is as important as weapon and body size for territorial ownership in a lekking hummingbird," *Scientific Reports* 8, no. 2001 (2018): doi:10.1038/s41598-018-20441-x.

H. J. Batcheller, "Interspecific information use by army-ant-following birds," *Auk* 134, no. 1 (2017): 247–55.

J. C. Bednarz, "Cooperative hunting Harris' hawks (*Parabuteo unicinctus*)," *Science* 239, no. 4847 (1988): 1525–27.

S. Boinski and P. E. Scott, "Association of birds with monkeys in Costa Rica," *Biotropica* 20, no. 2 (1988): 136–43.

J. Chaves-Campos, "Ant colony tracking in the obligate army ant-following antbird *Phaenostictus mcleannani*," *Journal of Ornithology* 152, no. 2 (2011): 497–504.

J. Chaves-Campos, "Localization of army-ant swarms by ant-following birds on the Caribbean slope of Costa Rica: Following the vocalization of antbirds to find the swarms," *Ornitologia Neotropical* 14, no. 3 (2003): 289–94.

J. Chaves-Campos et al., "The effect of local dominance and reciprocal tolerance on feeding aggregations of ocellated antbirds," *Proceedings of the Royal Society B* 276, no. 1675 (2009): 3995–4001.

L. G. Cheke and N. S. Clayton, "Mental time travel in animals," *Wiley Interdisciplinary Reviews: Cognitive Science* 1, no. 6 (2010): doi.org/10.1002/wcs.59.

N. S. Clayton and A. Dickinson, "Episodic-like memory during cache recovery by scrub-jays," *Nature* 395, no. 6699 (1998): 272–78.

M. A. W. Hornsby et al., "Wild hummingbirds can use the geometry of a flower array," *Behavioural Processes* 139 (2017): 33–37.

C. J. Logan et al., "A case of mental time travel in ant-following birds?," *Behavioral Ecology* 22, no. 6 (2011): 1149–53.

A. E. Martínez et al., "Social information cascades influence the formation of mixed-species foraging aggregations of ant-following birds in the Neotropics," *Animal Behaviour* 135 (2018): 25–35.

S. O'Donnell, "Evidence for facilitation among avian army-ant attendants: Specialization and species associations across elevations," *Biotropica* 49, no. 5 (2017): 665–74: doi.org/10.1111/btp.12452.

S. O'Donnell et al., "Specializations of birds that attend army ant raids: An ecological approach to cognitive and behavioral studies," *Behavioural Processes* 91 (2012): 267–74.

F. Otto, "5 Things to Know About Being Bitten by a Viper," *Drexel University News Blog*, December 21, 2015.

D. J. Pritchard and S. D. Healy, "Taking an insect-inspired approach to bird navigation," *Learning and Behavior* 46, no. 1 (2018): 7–22.

T. Suddendorf and M. C. Corballis, "The evolution of foresight: What is mental time travel, and is it unique to humans?," *Behavioral and Brain Sciences* 30, no. 3 (2007): 299–351.

M. B. Swartz, "Bivouac checking, a novel behavior distinguishing obligate from opportunistic species of army-ant-following birds," *Condor: Ornithological Applications* 103, no. 3 (2001): 629–33.

J. Tooby and I. DeVore, "The Reconstruction of Hominid Behavioral Evolution Through Strategic Modeling," in *The Evolution of Human Behavior: Primate Models*, ed. W. G. Kinzey (Albany: State University of New York Press, 1987): 183–237.

J. M. Touchton and J. N. M. Smith, "Species loss, delayed numerical responses, and functional compensation in an antbird guild," *Ecology* 92, no. 5 (2011): 1126–36.

J. M. Touchton and M. Wikelski, "Ecological opportunity leads to the emergence of an alternative behavioural phenotype in a tropical bird," *Journal of Animal Ecology* 84, no. 4 (2015): 1041–49.

E. Tulving, "Episodic memory: From mind to brain," *Annual Review of Psychology* 53 (2002): 1–25.

E. O. Willis, *The Behavior of Ocellated Antbirds* (Washington, DC: Smithsonian Institution Press, 1973).

E. O. Wilson, *A Window on Eternity, A Biologist's Walk Through Gorongosa National Park* (New York: Simon & Schuster, 2014).

E. O. Willis and Y. Oniki, "Birds and army ants," *Annual Review of Ecology and Systematics* 9 (1978): 243–63.

P. H. Wrege et al., "Antbirds parasitize foraging army ants," *Ecology* 86, no. 3 (2005): 555–59.

PLAY

CHAPTER SEVEN: BIRDS OF PLAY

J. E. C. Adriaense et al., "Negative emotional contagion and cognitive bias in common ravens (*Corvus corax*)," *PNAS* 116, no. 23 (2019): 11547–52.

A. C. Bent, *Life Histories of North American Jays, Crows, and Titmice* (Mineola, NY: Dover, 1964).

K. Bobrowicz and M. Osvath, "Cognition in the fast lane: Ravens' gazes are half as short as humans' when choosing objects," *Animal Behavior and Cognition* 6, no. 2 (2019): 81–97.

A. Bond and J. Diamond, *Thinking Like a Parrot: Perspectives from the Wild* (Chicago: University of Chicago Press, 2019).

T. Bugnyar et al., "Ravens attribute visual access to unseen competitors," *Nature Communications* 7, no. 10506 (2016): doi: 10.1038/ncomms10506.

T. Bugnyar et al., "Ravens judge competitors through experience with play caching," *Current Biology* 17, no. 20 (2007): 1804–8.

G. M. Burghardt, "Defining and Recognizing Play," article in the *Oxford Handbook of the Development of Play* (Oxford: Oxford University Press, 2012).

N. J. Emery and N. S. Clayton, "Do birds have the capacity for fun?," *Current Biology* 25, no. 1 (2015): R16–R20.

M. S. Ficken, "Avian play," *Auk* 94 (1977): 573–82.

E. H. Forbush and John Bichard May, *A Natural History of American Birds of Eastern and Central North America* (New York: Bramhall House, 1955).

K. Groos, *The Play of Animals*, trans. Elizabeth L. Baldwin (New York: D. Appleton and Company, 1898).

E. Gwinner, "Über einige Bewegungsspiele des Kolkraben," *Zeitschrift für Tierpsychol* 23 (1966): 28–36.

B. Heinrich, *Mind of the Raven* (NY: HarperCollins, 1999).

B. Heinrich, "Why do ravens fear their food?", *The Condor* 90 (1988): 950–52.

J. Hutto, *Illumination in the Flatwoods: A Season with the Wild Turkey* (Guilford, CT: Lyons and Burford, 1995).

I. Jacobs et al., "Object caching in corvids: Incidence and significance," *Behavioural Processes* 102 (2014): 25–32.

C. Kabadayi and M. Osvath, "Ravens parallel great apes in flexible planning for tool-use and bartering," *Science* 357, no. 6347 (2017): 202–4.

M. L. Lambert et al., "Birds of a feather? Parrot and corvid cognition compared," *Behaviour* 156, nos. 5–8 (2018): 508–94.

R. Miller et al., "Differences in exploration behaviour in common ravens and carrion crows during development and across social context," *Behavioral Ecology and Sociobiology* 69, no. 7 (2015): 1209–20.

E. P. Moreno-Jiménez et al., "Adult hippocampal neurogenesis is abundant in neurologically healthy subjects and drops sharply in patients with Alzheimer's disease," *Nature Medicine* 25, no. 4 (2019): 554–60.

M. Osvath and M. Sima, "Sub-adult ravens synchronize their play: A case of emotional contagion?" *Animal Behavior and Cognition* 1, no. 2 (2014): 197–205.

M. Osvath et al., "An exploration of play behaviors in raven nestlings," *Animal Behavior and Cognition* 1, no. 2 (2014): 157–65.

S. M. Pellis et al., "Is play a behavior system, and, if so, what kind?," *Behavioural Processes* 160 (2019): 1–9.

L. Riters et al., "Song practice as a rewarding form of play in songbirds," *Behavioural Processes* 163 (2017): doi.org/10.1016/j.beproc.2017.10.002.

S. M. Smith, "The behavior and vocalizations of young turquoise-browed motmots," *Biotropica* 9, no. 2 (1977): 127–30.

M. Spinka et al., "Mammalian play: Training for the unexpected," *Quarterly Review of Biology* 76, no. 2 (2001): 141–68.

D. Van Vuren, "Aerobatic rolls by ravens on Santa Cruz Island, California," *Auk* 101, no. 3 (1984): 620–21.

CHAPTER EIGHT: CLOWNS OF THE MOUNTAINS

J. Diamond and A. B. Bond, *Kea, Bird of Paradox: The Evolution and Behavior of a New Zealand Parrot* (Oakland: University of California Press, 1999).

G. K. Gajdon et al., "What a parrot's mind adds to play: The urge to produce novelty fosters tool use acquisition in kea," *Open Journal of Animal Sciences* 4 (2014): 51–58.

M. Goodman et al., "Habitual tool innovated by free-living New Zealand kea," *Sci Rep.* 8, no. 1 (2018): 13935: doi:10.1038/s41598-018-32363-9.

M. Heaney et al., "Keas perform similarly to chimpanzees and elephants when solving collaborative tasks," *PLOS ONE* 12, no. 2 (2017): e0169799: doi.org/10.1371/journal.pone.0169799.

J. R. Jackson, "Keas at Arthurs Pass," *Notornis* 9, no. 2 (1960): 39–58.

G. Marriner, *The Kea: A New Zealand Problem* (Christchurch, NZ: Marriner Bros., 1908).

M. O'Hara et al., "Kea Logics: How These Birds Solve Difficult Problems and Outsmart Researchers," in *Logic and Sensibility*, ed. S. Watanabe, (Keio, Japan: Center for Advanced Research on Logic and Sensibility, 2012).

J. Panksepp, "Beyond a joke: From animal laughter to human joy?," *Science* 308, no. 5718 (2005): 62–63.

S. M. Pellis et al., "The function of play in the development of the social brain," *American Journal of Play* 2, no. 3 (2010): 278–97.

R. Schwing, "Scavenging behavior of kea (*Nestor notabilis*)," *Notornis* 57, no. 2 (2010): 98–99.

R. Schwing et al., "Kea (*Nestor notabilis*) decide early when to wait in food exchange task," *Journal of Comparative Psychology* 131, no. 4 (2017): 269–76.

R. Schwing et al., "Positive emotional contagion in a New Zealand parrot," *Current Biology* 27, no. 6 (2017): R213–R214.

R. Schwing et al., "Vocal repertoire of the New Zealand kea parrot *Nestor notabilis*," *Current Zoology* 58, no. 5 (2012): 727–40.

A. Wein et al., "Picture—object recognition in kea (*Nestor notabilis*)," *Ethology* 121, no. 11 (2015): 1059–70.

LOVE

CHAPTER NINE: SEX

P. Abbassi and N. T. Burley, "Nice guys finish last: Same-sex sexual behavior and pairing success in male budgerigars," *Behavioral Ecology* 23, no. 4 (2012): 775–82.

D. G. Ainley, "Displays of Adélie Penguins: A Re-interpretation," (1974) in *The Biology of Penguins*, ed. B. Stonehouse, (London: Macmillan, 1975), 503–34.

N. W. Bailey and M. Zuk, "Same-sex sexual behavior and evolution," *Trends in Ecology and Evolution* 24, no. 8 (2009): 439–46.

Y. Ben Mocha and S. Pika, "Intentional presentation of objects in cooperatively breeding Arabian babblers (*Turdoides squamiceps*), *Frontiers in Ecology and Evolution* (2019): doi.org/10.3389/fevo .2019.00087.

Y. Ben Mocha et al., "Why hide? Concealed sex in dominant Arabian babblers (*Turdoides squamiceps*) in the wild," *Evolution and Human Behavior* 39 (2018): 575–82.

T. Birkhead, "Uncovered: The Secret Sex Life of Birds," BirdLife International, February 13, 2018.

P. L. R. Brennan, "The Hidden Side of Sex," *Scientist*, July 1, 2014.

P. L. R. Brennan and R. Prum, "The limits of sexual conflict in the narrow sense: New insights from waterfowl biology," *Philosophical Transactions of the Royal Society B* 367, no. 1600 (2012): 2324–38.

G. M. Levick, *Antarctic Penguins: A Study of Their Social Habits* (London: William Heinemann, 1914).

G. R. MacFarlane et al., "Homosexual behaviour in birds: Frequency of expression is related to parental care disparity between the sexes," *Animal Behaviour* 80, no. 3 (2010): 375–90.

D. MacLeod, "Necrophilia among ducks ruffles research feathers," *The Guardian*, March 8, 2005.

C. W. Moeliker, "The first case of homosexual necrophilia in the mallard *Anas platyrhynchos* (Aves: Anatidae)," *DEINSEA* 8 (2001): 243–47.

A. P. Møller, "Copulation behaviour in the goshawk, *Accipiter gentilis*," *Animal Behaviour* 35, no. 3 (1987): 755–63.

N. Ota, "Are the neural mechanisms shared between singing and dancing in Blue-Capped Cordon-Bleu Finches?", Presentation at the International Ornithological Congress 2018.

N. Ota et al., "Tap dancing birds: the multimodal mutual courtship display of males and females in a socially monogamous songbird," *Scientific Reports* 5 (16614) (2015).

D. G. D. Russell et al., "Dr. George Murray Levick (1876–1956): Unpublished notes on the sexual habits of the Adélie penguin," *Polar Record* 48, no. 4 (2012): 387–393.

K. Swift and J. M. Marzluff, "Occurrence and variability of tactile interactions between wild American crows and dead conspecifics," *Philosophical Transactions of the Royal Society B: Biological Sciences* 373, no. 1754 (2018): 20170259.

CHAPTER TEN: WILD WOOING

F. J. Aznar and M. Ibáñez-Agulleiro, "The function of stones in nest building: The case of black wheatear (*Oenanthe leucura*) revisited," *Avian Biology Research* 1 (2016): 3–12.

A. C. Bent, *Life Histories of North American Nuthatches, Wrens, Thrashers and Their Allies* (Washington, DC: Smithsonian, 1948).

N. J. Boogert et al., "Mate choice for cognitive traits: A review of the evidence in nonhuman vertebrates," *Behavioral Ecology* 22, no. 3 (2011): 447–59.

E. H. DuVal, "Cooperative display and lekking behavior of the lance-tailed manakin (*Chiroxiphia lanceolata*)," *Auk* 124, no. 4 (2007): 1168–85.

E. H. DuVal, "Female mate fidelity in a lek mating system and its implications for the evolution of cooperative lekking behavior," *American Naturalist* 181, no. 2 (2013): 213–22.

R. Heinsohn et al., "Tool-assisted rhythmic drumming in palm cockatoos shares key elements of human instrumental music," *Science Advances* 3, no. 6 (2017): e1602399.

B. G. Hogan and M. C. Stoddard, "Synchronization of speed, sound and iridescent color in a hummingbird aerial courtship dive," *Nature Communications* 9, no. 1 (2018): 5260.

M. G. Lockley et al., "Theropod courtship: Large scale physical evidence of display arenas and avianlike scrape ceremony behaviour by Cretaceous dinosaurs," *Scientific Reports* 6 (2016): 18952.

E. A. Marks et al., "Ecstatic display calls of the Adélie penguin honestly predict male condition and breeding success," *Behaviour* 147, no. 2 (2010): 165-184.

J. Moreno et al., "The function of stone carrying in the black wheateater, *Oenanthe leucura*" *Animal Behaviour* 47, no. 6 (1994): 1297–1309.

R. G. Prum, *The Evolution of Beauty: How Darwin's Forgotten Theory of Mate Choice Shapes the Animal World—and Us* (New York: Doubleday, 2017).

M. J. Ryan, *A Taste for the Beautiful: The Evolution of Attraction* (Princeton, NJ: Princeton University Press, 2018).

M. J. Ryan and M. E. Cummings, "Perceptual biases and mate choice," *Annual Review of Ecology, Evolution, and Systematics* 44 (2013): 437–59.

H. M. Schaefer and G. D. Ruxton, "Signal diversity, sexual selection, and speciation," *Annual Review of Ecology, Evolution, and Systematics* 46 (2015): 573–92.

E. Scholes, "Courtship ethology of Carola's parotia (*Parotia Carolae*)," *Auk* 123, no. 4 (2006): 967–90.

E. J. Scholes et al., "Visual and acoustic components of courtship in the bird-of-paradise genus *Astrapia* (Aves: Paradisaeidae)," *PeerJ* 5 (2017): e3987.

CHAPTER ELEVEN: BRAIN TEASERS

M. Araya-Salas et al., "Spatial memory is as important as weapon and body size for territorial ownership in a lekking hummingbird," *Scientific Reports* 8, no. 2001 (2018): doi:10.1038/s41598-018-20441-x.

G. Borgia, "Complex male display and female choice in the spotted bowerbird: Specialized functions for different bower decorations," *Animal Behaviour* 49, no. 5 (1995): 1291–301.

G. Borgia and J. Keagy, "Cognitively Driven Co-option and the Evolution of Complex Sexual Displays in Bowerbirds," in *Animal Signaling and Function: An Integrative Approach*, ed. D. J. Irschick et al. (Hoboken, NJ: John Wiley and Sons, 2015), 75–109.

G. Borgia and D. C. Presgraves, "Coevolution of elaborate male display traits in the spotted bowerbird: An experimental test of the threat reduction hypothesis," *Animal Behaviour* 56, no. 5 (1998): 1121–28.

J. Chen et al., "Problem-solving males become more attractive to female budgerigars," *Science* 363, no. 6423 (2019): 166–67.

A. H. Chisholm, *Bird Wonders of Australia* (East Lansing: Michigan State University Press, 1958).

A. Cockburn, "Can't See the 'Hood for the Trees: Phylogenetic and Ecological Pattern in Cooperative Breeding in Birds," plenary lecture, IOC 2018.

A. Cockburn et al., "Superb fairy-wren males aggregate into hidden leks to solicit extragroup fertilizations before dawn," *Behavioral Ecology* 20, no. 3 (2009): 501–10.

A. H. Dalziell and A. Cockburn, "Dawn song in superb fairy-wrens: A bird that seeks extrapair copulations during the dawn chorus," *Animal Behaviour* 75, no. 2 (2008): 489–500.

J. M. Diamond, "Bower building and decoration by the bowerbird *Amblyornis inornatus*," *Ethology* 74, no. 3 (1987): 177–204.

J. M. Diamond, "Evolution of bowerbirds' bowers: Animal origins of the aesthetic sense," *Nature* 297 (1982): 99–102.

J. Keagy et al., "Cognitive ability and the evolution of multiple behavioral display traits," *Behavioral Ecology* 23, no. 2 (2012): 448–56.

L. A. Kelley and J. A. Endler, "How do great bowerbirds create forced perspective illusions?," *Royal Society Open Science* 4, no. 1 (2017): 160661.

D. Lack, *Ecological Adaptations for Breeding in Birds* (London: Chapman and Hall, 1968).

J. R. Madden, "Do bowerbirds exhibit cultures?," *Animal Cognition* 11, no. 1 (2008): 1–12.

D. J. Pritchard and S. D. Healy, "Taking an insect-inspired approach to bird navigation," *Learning & Behavior* 46, no. 1 (2018): 7–22.

D. J. Pritchard et al., "Wild hummingbirds require a consistent view of landmarks to pinpoint a goal location," *Animal Behaviour* 137 (2018): 83–94.

D. J. Pritchard et al., "Wild rufous hummingbirds use local landmarks to return to rewarded locations," *Behavioural Processes* 122 (2016): 59–66.

PARENT

CHAPTER TWELVE: FREE-RANGE PARENTING

M. AlRashidi et al., "The influence of a hot environment on parental cooperation of a ground-nesting shorebird, the Kentish plover *Charadrius alexandrines*," *Frontiers in Zoology* 7 (2010): 1–10.

I. E. Bailey et al., "Image analysis of weaverbird nests reveals signature weave patterns," *Royal Society Open Science* 2, no. 6 (2015): 150074.

R. Biancalana, "Breeding biology of the sooty swift *Cypseloides fumigatus* in São Paulo, Brazil," *Wilson Journal of Ornithology* 127, no. 3 (2015): 402–10.

A. J. Breen et al., "What can nest-building birds teach us?," *Comparative Cognition and Behavior Reviews* 11 (2016): 83–102.

B. L. Campbell et al., "Behavioural plasticity under a changing climate; how an experimental local climate affects the nest construction of the zebra finch *Taeniopygia guttata*," *Journal of Avian Biology* 49, no. 4 (2018): doi.org/10.1111/jav.01717.

A. Cockburn, "Prevalence of different modes of parental care in birds," *Proceedings B: Biological Sciences* 273, no. 1592 (2006): 1375–83.

A. Göth and D. Jones, "Ontogeny of social behaviour in the megapode Australian brush-turkey (Alectura lathami)," *Journal of Comparative Psychology* 117, no. 1 (2003): 36–43.

H. F. Greeney et al., "Trait-mediated trophic cascade creates enemy-free space for nesting hummingbirds," *Science Advances* 1, no. 8 (2015): e1500310.

L. M. Guillette and S. D. Healy, "Nest building, the forgotten behavior," *Current Opinion in Behavioural Sciences* 6 (2015): 90–96.

L. M. Guillette et al., "Social learning in nest-building birds: A role for familiarity," *Proceedings of the Royal Society B: Biological Sciences* 283, no. 1827 (2016): 20152685.

Z. J. Hall et al., "Neural correlates of nesting behavior in zebra finches (*Taeniopygia guttata*)," *Behavioural Brain Research* 264, no. 100 (2014): 26–33.

Z. J. Hall et al., "A role for nonapeptides and dopamine in nest-building behaviour," *Journal of Neuroendocrinology* 27, no. 2 (2015): 158–65.

M. R. Halley and C. M. Heckscher, "Interspecific parental care by a wood thrush (*Hylocichla mustelina*) at a nest of the veery (*Catharus fuscescens*)," *Wilson Journal of Ornithology* 125, no. 4 (2013): 823–28.

R. Heinsohn et al., "Adaptive sex ratio adjustments via sex-specific infanticide in a bird," *Current Biology* 21, no. 20 (2011): 1744–47.

D. N. Jones et al., *The Megapodes* (Oxford: Oxford University Press, 1995).

D. N. Jones et al., "Presence and distribution of Australian brush-turkeys in the greater Brisbane region," *Sunbird* 34, no. 1 (2004): 1–9.

D. N. Jones, "Living with a dangerous neighbor: Australian magpies in a suburban environment," *Proceedings 4th International Urban Wildlife Symposium*, ed. Shaw et al. (2004).

D. N. Jones, "Reproduction Without Parenthood: Individual Behaviour of Male, Female and Juvenile Australian Brush-Turkeys," in *Animal Societies: Individuals, Interaction and Organisation*, eds. Peter J. Jarman and Andrew Rossiter (Kyoto: Kyoto University Press, 1994): 135–146.

J. J. Price and S. C. Griffith, "Open cup nests evolved from roofed nests in the early passerines," *Proceedings of the Royal Society B* 284, no. 1848 (2017): doi.org/10.1098/rspb.2016.2708.

D. R. Rubenstein, "Superb starlings: cooperation and conflict in an unpredictable environment," in *Cooperative Breeding in Vertebrates*, eds. W. D. Koenig and J. L. Dickinson (Cambridge: Cambridge University Press, 2016): 181–96.

M. M. Shy, "Interspecific feeding among birds: A review," *Journal of Field Ornithology* 53, no. 4 (1982): 370–93.

M. C. Stoddard et al., "Avian egg shape: Form, function and evolution," *Science* 356, no. 6344 (2017): 1249–54.

M. C. Stoddard et al., "Evolution of avian egg shape: Underlying mechanisms and the importance of taxonomic scale," *Ibis* 161 (2019): 922–25: doi.org/10.1111/ibi.12755.

R. E. van Dijk et al., "Nest desertion is not predicted by cuckoldry in the Eurasian penduline tit," *Behavioral Ecology and Sociobiology* 64, no. 9 (2010): 1425–35.

K. van Vuuren et al., "'Vicious, aggressive bird stalks cyclist': The Australian magpie (*Cracticus tibicen*) in the news," *Animals* 6, no 5 (2016): 29.

CHAPTER THIRTEEN: THE WORLD'S BEST BIRDWATCHERS

D. E. Blasi et al., "Sound–meaning association biases evidenced across thousands of languages," *PNAS* 113, no. 39 (2016): 10818–23.

W. E. Feeney, "Evidence of Adaptations and Counter-Adaptations Before the Parasite Lays Its Egg: The Frontline of the Arms Race," in *Avian Brood Parasitism: Behaviour, Ecology, Evolution and Coevolution*, ed. M. Soler (New York: Springer, 2017): 307–24.

W. E. Feeney, "'Jack-of-all-trades' egg mimicry in the brood parasitic Horsfield's bronze-cuckoo?" *Behavioral Ecology* 25, no. 6 (2014): 1365–73.

W. E. Feeney and N. E. Langmore, "Social learning of a brood parasite by its host," *Biology Letters* 9, no. 4 (2013): doi.org/10.1098/rsbl.2013.0443.

W. E. Feeney and N. E. Langmore, "Superb fairy-wrens, *Malurus cyaneus*, increase vigilance near their nest with the perceived risk of brood parasitism," *Auk* 132, no. 2 (2015): 359–64.

W. E. Feeney et al., "Advances in the study of coevolution between avian brood parasites and their hosts," *Annual Review of Ecology, Evolution, and Systematics* 45 (2014): 227–46.

W. E. Feeney et al., "Evidence for aggressive mimicry in an adult brood parasitic bird, and generalised defences in its host," *Proceedings of the Royal Society B: Biological Sciences* 282, no. 1810 (2015): 20150795.

W. E. Feeney et al., "The frontline of avian brood parasite-host coevolution," *Animal Behaviour* 84 (2012): 3–12.

M. F. Guigueno et al., "Female cowbirds have more accurate spatial memory than males," *Biology Letters* 10, no. 2 (2014): 20140026.

H. A. Isack and H.-U. Reyer, "Honeyguides and honey gatherers: Interspecific communication in a symbiotic relationship," *Science* 243, no. 4896 (1989): 1343–46.

R. M. Kilner and N. E. Langmore, "Cuckoos versus hosts in insects and birds: Adaptations, counteradaptations and outcomes," *Biological Reviews* 86, no. 4 (2011): 836–52.

N. E. Langmore et al., "Escalation of a coevolutionary arms race through host rejection of brood parasitic young," *Nature* 422, no. 6928 (2003): 157–60.

N. E. Langmore et al., "Learned recognition of brood parasitic cuckoos in the superb fairy-wren, *Malurus cyaneus*," *Behavioral Ecology* 23, no. 4 (2012): 798–805.

N. E. Langmore et al., "Visual mimicry of host nestlings by cuckoos," *Proceedings of the Royal Society B: Biological Sciences* 278, no. 1717 (2011): 2455–63.

N. E. Langmore et al., "Socially acquired host-specific mimicry and the evolution of host races in Horsfield's bronze-cuckoo *Chalcites basalis*," *Evolution* 62, no. 7 (2008): 1689–99.

W. Liang, "Crafty cuckoo calls," *Nature Ecology & Evolution* 1, no. 10 (2017): 1427–28.

M. I. M. Louder et al., "An acoustic password enhances auditory learning in juvenile brood parasitic cowbirds," *Current Biology* (2019): doi.org/10.1016/j.cub.2019.09.046.

M. I. M. Louder et al., "A generalist brood parasite modifies use of a host in response to reproductive success," *Proceedings of the Royal Society B: Biological Sciences* 282, no. 1814 (2015): doi.org/10.1098/rspb.2015.1615.

D. Parejo and J. M. Avilés, "Do avian brood parasites eavesdrop on heterospecific sexual signals revealing host quality? A review of the evidence," *Animal Cognition* 10, no. 2 (2007): 81–88.

E. Pennisi, "Wild bird comes when honey hunters call for help," *Science* 353, no. 6297 (2016): 335.

G. A. Ranger, "On three species of honey-guide; the greater (*Indicator indicator*), the lesser (*Indicator minor*) and the scaly-throated (*Indicator variegatus*)," *Ostrich* 26, no. 2 (1955): 70–87.

J. M. Rojas Ripari et al., "Innate development of acoustic signals for host parent–offspring recognition in the brood-parasitic screaming cowbird *Molothrus rufoaxillaris*," *Ibis* 161, no. 4 (2018): 717–19.

C. N. Spottiswoode and J. Koorevaar, "A stab in the dark: Chick killing by brood parasitic honeyguides," *Biology Letters* 8, no. 2 (2012): 241–44.

C. N. Spottiswoode and M. N. Stevens, "Host-parasite arms races and rapid changes in bird egg appearance," *American Naturalist* 179, no. 5 (2012): 633–48.

C. N. Spottiswoode et al., "Reciprocal signaling in honeyguide-human mutualism," *Science* 353, no. 6297 (2016): 387–89.

M. Stevens, "Bird brood parasitism," *Current Biology* 23, no. 20 (2013): R909–R913.

M. C. Stoddard and M. E. Hauber, "Colour, vision and coevolution in avian brood parasitism," *Philosophical Transactions of the Royal Society B: Biological Sciences* 372, no. 1724 (2017): 20160339.

M. C. Stoddard et al., "Higher-level pattern features provide additional information to birds when recognizing and rejecting parasitic eggs," *Philosophical Transactions of the Royal Society B: Biological Sciences* 374, no. 1769 (2019): 20180197.

CHAPTER FOURTEEN: A CHILDCARE COOPERATIVE OF WITCHES AND WATER BOILERS

B. J. Ashton et al., "Cognitive performance is linked to group size and affects fitness in Australian magpies," *Nature* 554, no. 7692 (2018): 364–67.

B. J. Ashton et al., "An intraspecific appraisal of the social intelligence hypothesis," *Philosophical Transactions of the Royal Society B: Biological Sciences* 373, no. 1756 (2017): doi/10.1098/rstb.2017.0288.

A. Carr, *The Windward Road: Adventures of a Naturalist on Remote Caribbean Shores* (New York: Knopf, 1956).

C. K. Cornwallis, "Cooperative breeding and the evolutionary coexistence of helper and nonhelper strategies," *Proceedings of the National Academy of Sciences* 115, no. 8 (2018): 1684–86.

W. D. Hamilton, "The genetical evolution of social behaviour. II," *Journal of Theoretical Biology* 7 (1964): 17–52.

W. D. Koenig, "What drives cooperative breeding?," *PLOS Biology* 15, no. 6 (2017): e2002965.

W. D. Koenig and R. L. Mumme, "The great egg-demolition derby," *Natural History* 106, no. 5 (1997): 32–37.

R. L. Mumme et al., "Costs and benefits of joint nesting in the acorn woodpecker," *American Naturalist* 131, no. 5 (1988): 654–77.

A. Ridley et al., "The cost of being alone: The fate of floaters in a population of cooperatively breeding pied babblers *Turdoides bicolor*," *Journal of Avian Biology* 3, no. 4 (2008): 389–92.

C. Riehl, "Infanticide and within-clutch competition select for reproductive synchrony in a cooperative bird," *Evolution* 70, no. 8 (2016): 1760–69.

C. Riehl and M. J. Strong, "Social living without kin discrimination: Experimental evidence from a communally breeding bird," *Behavioral Ecology and Sociobiology* 69, no. 8 (2015): 1293–99.

C. Riehl and M. J. Strong, "Stable social relationships between unrelated females increase individual fitness in a cooperative bird," *Proceedings of the Royal Society B: Biological Sciences* 285, no. 1876 (2018): doi/10.1098/rspb.2018.0130.

C. Riehl et al., "Inferential reasoning and egg rejection in a cooperatively breeding cuckoo," *Animal Cognition* 18, no. 1 (2015): 75–82.

M. Taborsky et al., "The evolution of cooperation based on direct fitness benefits," *Philosophical Transactions of the Royal Society B: Biological Sciences.* 371, no. 1687 (2016): 20140474.

A. Thornton and K. McAuliffe, "Cognitive consequences of cooperative breeding? A critical appraisal," *Journal of Zoology* 295, no. 1 (2015): 12–22.

K. Wojczulanis-Jakubas et al., "Seabird parents provision their chick in a coordinated manner," *PLOS ONE* 13, no. 1 (2018): e0189969.

A LAST WORD

D. E. Bowler, "Long-term declines of European insectivorous bird populations and potential causes," *Conservation Biology* 33, no. 5 (2019): 1120–30.

M. L. Eng et al., "A neonicotinoid insecticide reduces fueling and delays migration in songbirds," *Science* 365, no. 6458 (2019): 1177–80.

C. M. Heckscher, "A nearctic-neotropical migratory songbird's nesting phenology and clutch size are predictors of accumulated cyclone energy," *Scientific Reports* 8, no. 9899 (2018): doi:10.1038 /s41598-018-28302-3.

S. Jenouvrier et al., "The Paris agreement objectives will likely halt future declines of emperor penguins," *Global Change Biology* (2019): doi/10.1111/gcb.14864.

C. W. Rettenmeyer et al., "The largest animal association centered on one species: The army ant *Eciton burchellii* and its more than 300 associates," *Insectes Sociaux* 58, no. 3 (2011): 281–92.

L. Robin and R. Heinsohn, *Boom & Bust: Bird Stories for a Dry Country* (Clayton, Victoria, Australia: Csiro Publishing, 2009).

K. V. Rosenberg et al., "Decline of the North American avifauna," *Science* 366, no. 6461 (2019): eaaw1313.

Index

THE GENIUS OF BIRDS

Birds are astonishingly intelligent creatures. According to revolutionary new research, some birds rival primates and even humans in their remarkable mental prowess. In *The Genius of Birds*, acclaimed author Jennifer Ackerman tours the globe to learn more about the recently uncovered genius of birds, and dives deep into the latest findings about the bird brain itself. At once personal yet scientific, richly informative and beautifully written, *The Genius of Birds* celebrates the triumphs of these surprising and fiercely intelligent creatures.

BIRDS BY THE SHORE

Observing the Natural Life of the Atlantic Coast

For three years, Jennifer Ackerman lived in the small coastal town of Lewes, Delaware, in the sort of blue-water, white-sand landscape that draws summer crowds up and down the eastern seaboard. *Birds by the Shore* is a book about discovering the natural life at the ocean's edge: the habits of shorebirds and seabirds, the movement of sand and water. Against this landscape's rhythms, Ackerman revisits her own history—her mother's death, her father's illness, and her hopes to have children of her own.

 PENGUIN BOOKS